Parametric Estimates by the Monte Carlo Method

Parametric Estimates by the Monte Carlo Method

G.A. Mikhailov

///VSP///
Utrecht, The Netherlands, 1999

VSP BV
P.O. Box 346
3700 AH Zeist
The Netherlands

Tel: +31 30 692 5790
Fax: +31 30 693 2081
E-mail: vsppub@compuserve.com
Home Page: http://www.vsppub.com

First published in 1999

ISBN 90-6764-297-5

Printed in The Netherlands by Ridderprint bv, Ridderkerk.

Contents

PREFACE

This monograph is devoted to the further development of parametric weight Monte Carlo estimates for solving linear and nonlinear integral equations and boundary value problems for equations with partial derivatives. These estimates are related to special auxiliary weight factors which explicitly depend on parameters while simulated Markov chains do not depend on them. Using these estimates, it is possible to construct effective Monte Carlo methods for calculating parametric and coordinate multiple derivatives of solutions and linear functionals. On this basis it is possible to estimate the critical values of parameters which sometimes coincide with the main eigenvalues of the corresponding operators.

It is well known that Monte Carlo methods are especially effective for solving problems with stochastic parameters. In a special chapter of the monograph, the new Monte Carlo methods for solving stochastic radiation transfer problems are presented, including the estimation of probabilistic moments of the corresponding critical parameters. In a sense, the monograph can be regarded as the sequel of the previous anthor's monograph "New Monte Carlo Methods with Estimating Derivatives" (VSP, 1995). Some material from this monograph is repeated here with necessary modifications, but mainly the material herein is new and contains many references to the previous monograph. The first chapter of the monograph is in fact a short manual on the theory of weight Monte Carlo methods including the improved grounds of the parametric differentiation and the additional randomization of weight estimates. Some proofs in an improved form are stated here which correspond to a new university lecture course.

The second chapter contains new material on solving boundary value problems with complex parameters, mixed problems to parabolic equations, boundary value problems of the second and third kinds, and some improved techniques related to vector and nonlinear Helmholtz equations. There is a section on estimation of the main eigenvalue of the Laplace operator by calculating multiple parametric derivatives when solving the special Helmholtz equation.

Special attention is paid to the foundation and optimization of the known global 'walk on grid' method for solving the Helmholtz difference equation. It seems that this method is appropriate for solving any elliptic difference equation and it is more effective than the 'walk on spheres' method for the global estimation of a solution. However, the walk-on-spheres method is preferable if it is necessary to estimate parametric or coordinate derivatives of a solution.

In the third chapter, random media models related to homogeneous point fluxes are used. For these models new asymptotic estimates of the radiation transfer are obtained and the comparative Monte Carlo calculations are done. It turned out that the corresponding weight estimates can be partly averaged with respect to values of a random density. Using these new estimates the influence of a model dimension on the radiation transfer was estimated.

In the special extended section 3.4, probabilistic moments of the time constant and the effective multiplication factor in a stochastic medium with scattering, absorption and fission are estimated. For this purpose, corresponding formulae for averaged constants of integro-differential transfer equations are constructed, and the diffusion approximation and simple probabilistic models, as well as a special iteration process realized with the Monte Carlo method, are used. Also considered is the possibility of estimating critical values of parameters of averaged solutions. Computational results for inform and two-layer balls with stochastic density values are presented.

Section 3.5 presents a new approach to constructing and substantiating weight 'path estimates' on the basis of introducing an artificial capture coefficient. Section 3.6 is devoted to the computation of parametric derivatives when solving problems of radiation transfer with polarization.

In the first appendix, the theoretical aspects of the familiar method of improving random number generators by modulo 1 summation are developed. There is considered the vector version of congruent summation which allows us to decrease the possible dependence of numbers by enhancing the property of uniformity in multidimensional distributions.

In the second appendix, algorithms of statistical modelling of chemical reactions are considered. The parametric condition of coexistence of steady states in chemical systems is derived and the corresponding estimates are constructed.

In the third appendix, one unsolved minimax problem related to optimizing parametric estimates is presented.

The work was supported by the Russian Basic Research Foundation (Grant No. 97-01-0855) and by Grant No. 3759 of the Scientific Programme "Universities of Russia–fundamental researches".

Chapter 1

Introduction: weight unbiased Monte Carlo estimates

1.1. Integral equations, linear functionals

Let us consider an integral equation of the second kind,

$$\varphi(x) = \int_X k(x', x)\varphi(x')\,dx' + f(x) \text{ or } \varphi = K\varphi + f, \tag{1.1}$$

where X is the n-dimensional Euclidean space, $f, \varphi \in L$, the Banach space of integrable functions. For many applications (for example in transfer theory), we put $L = L_1$, where

$$||f|| = \int_X |f(x)|\,dx, \qquad ||K|| \leq \text{vrai}\sup_x \int_X |k(x, x')|\,dx'.$$

The Monte Carlo algorithms constructed here and subsequently are based on representation of the solution of (1.1) by the Neumann series,

$$\varphi = \sum_{n=0}^{\infty} K^n f, \tag{1.2}$$

$$[K^n f](x) = \overbrace{\int \ldots \int}^{n} f(x_0)k(x_0, x_1)\ldots k(x_{n-1}, x)\,dx_0 \ldots dx_{n-1}.$$

It is easy to show that series (1.2) converges (in the norm) and the solution of (1.1) exists if $||K|| < 1$. But, for convergence of the Neumann series and existence of a solution, it is sufficient to require $||K^{n_0}|| < 1$ for some integer $n_0 \geq 1$. This follows from the relation $\varphi = K^{n_0}\varphi + K^{n_0-1}f + \ldots + Kf + f$, which is equivalent to (1.1). Let

$$\varphi^* = K^*\varphi^* + h \tag{1.3}$$

be the equation adjoint to (1.1), where $\varphi^*, h \in L^*, K^* \in [L^* \to L^*], L^*$ is the space conjugate to L, K^* is an operator conjugate to K.

Note that the condition : $||K^{n_0}|| < 1$, for some $n_0 \geq 1$, is equivalent to the inequality $\rho(K) < 1$, where the spectral radius $\rho(K)$ is determined by the relation:

$$\rho(K) = \rho(K^*) = \lim ||K^n||^{1/n} = \inf ||K^n||^{1/n}.$$

We recall that $|(f, h)| \leq ||f||_L \cdot ||h||_{L^*}$, where

$$(f, h) = \int_X f(x)h(x)\,dx, \qquad ||K|| = ||K^*||,$$

$$(Kf,h) = (f,K^*h), \quad [K^*h](x) = \int_X k(x,x')h(x')\,dx'.$$

By definition, L_∞ is a space conjugate to L_1, i.e. the space of bounded (almost everywhere) functions characterized by a norm

$$||h||_{L_\infty} = \text{vrai sup}\,|h(x)|, \quad x \in X.$$

In this chapter we consider Monte Carlo algorithms for estimating functionals of the form

$$I_h = (\varphi,h) = \sum_{n=0}^{\infty}(K^n f,h) = (f,\varphi^*). \tag{1.4}$$

The convergence of (1.4) is assured by the convergence (in norm) of the Neumann series.

In order to show that $(\varphi,h) = (f,\varphi^*)$, where $\varphi^* = K^*\varphi^*+h$, we multiply (1.1) by φ^* and (1.3) by φ and then compare these expressions, keeping in mind that $(K\varphi,\varphi^*) = (\varphi,K^*\varphi^*)$. So, the operator K is related to the kernel $k(x',x)$, and the conjugate operator K^* is related to the transposed kernel $k(x,x')$. However, for the sake of simplicity, operators with transposed-type kernels sometimes are denoted by symbols without asterisks, when only the operators of this type are used.

1.2. Terminating Markov chains

The stationary Markov chain is defined as the sequence of random points (states) $x_0, x_1, \ldots, x_n, \ldots$ such that the distribution of x_n is independent of all previous states except its immediate predecessor x_{n-1} or, more formally (for distribution densities)

$$\begin{aligned} P(x_n = x | x_{n-1} = x', \ldots, x_2 = s_2, x_1 = s_1) &= P(x_n = x | x_{n-1} = x') \\ &= r(x',x). \end{aligned}$$

The function $r(x',x)$ is called the transition probability density function (transition p. d. f.) and is sometimes denoted by $r(x' \to x)$. The distribution of the initial state is defined by the initial density $\pi(x)$. We may extend this definition by introducing a termination probability $p(x')$ at the point x' (or, at the transition $x' \to x$). The random number of the terminal state is denoted by N. Clearly, the Monte Carlo method deals only with Markov chains, having a finite number of steps with probability one. Moreover, it is usually supposed that the mean value $\mathrm{E}(N)$ is finite. Further, we give a sufficient condition for this. We emphasize again that the Markov chain is completely defined by the initial density $\pi(x)$, the transition p. d. f. $r(x',x)$ and the probability of termination $p(x')$.

Further, we also use the transition density $p(x',x) = r(x',x)[1-p(x')]$, which uniquely determines the functions $r(x',x)$ and $p(x')$, because

$$\int_X p(x',x)\,dx = 1 - p(x').$$

We use the notation : $B_p \in [L_1 \to L_1]$ is the operator with the kernel function $p(x',x)$. First, we obtain the probability of the event $\{N = n\}$:

$$\begin{aligned}
P(N = n) &= E\left[P(N = n|x_0,\dots,x_n)\right] = E\left[p(x_n) \times \prod_{k=0}^{n-1}(1 - p(x_k))\right] \\
&= \overbrace{\int_X \dots \int_X}^{n+1} \pi(x_0)p(x_n)\left[\prod_{k=0}^{n-1}(1 - p(x_k)) \times r(x_k, x_{k+1})\right] dx_0 \dots dx_n \\
&= \overbrace{\int_X \dots \int_X}^{n+1} \pi(x_0)p(x_n)\left[\prod_{k=0}^{n-1} p(x_k, x_{k+1})\right] dx_0 \dots dx_n \\
&= (K_p^n \pi, p).
\end{aligned} \tag{1.5}$$

To obtain $P(N \geq n)$, we must replace $p(x)$ with $\delta(x) \equiv 1$; hence $P(N \geq n) = (K_p \pi, \delta)$. Now

$$\{N = \infty\} \supset \{N \geq n\} \Rightarrow P(N = \infty) \leq P(N \geq n).$$

The relations just obtained show that $P(N > n) \to 0$ as $n \to \infty$; hence $P(N = \infty) = 0$, provided that the Neumann series for $f = B_p f + \pi$ converges. But, as was mentioned previously, for this, it is sufficient to require

$$||B_p^{n_0}|| < 1 \text{ for some } n_0, \text{ or } \rho(B_p) < 1. \tag{1.6}$$

Thus, if (1.6) holds, the Markov chain terminates after a finite number of steps with probability one. It is easy to show that (1.6) is a sufficient condition for $E(N)$ to be finite. Indeed,

$$\begin{aligned}
E(N) &= \sum_{n=0}^{\infty} n(B_p^n \pi, p) = \Big(\sum_{n=0}^{\infty} n B_p^n \pi, p\Big) \\
&= \Big(\sum_{n=1}^{\infty}\sum_{k=n}^{\infty} B_p^n, \pi\Big) = \Big(\sum_{n=1}^{\infty} B_p^n f, p\Big) = (f_p, p) < +\infty,
\end{aligned}$$

where $f = B_p f + \pi$, $f_p = B_p f_p + B_p f$.

By (1.6), we have $f, f_p \in L_1$, $(f_p, p) < +\infty$. In particular, $E(N) < +\infty$ if $p(x) \geq \varepsilon > 0$, because

$$||B_p|| \leq \sup_x \int_X r(x, x')[1 - p(x)]\, dx' \leq (1 - \varepsilon) \sup_x \int_X r(x, x')\, dx' = 1 - \varepsilon < 1.$$

So, $E(N) < +\infty$ and therefore $P(N = +\infty) = 0$, if (1.6) holds. In Section 1.3 we obtain these relations directly from the theorem about the biasedness of the main Monte Carlo estimate for the functional (φ, h).

1.3. Standard weight estimates in the Monte Carlo method, biasedness

1. Let $I_h = (\varphi, h)$ be the quantity that is to be estimated, where $\varphi = K\varphi + h$ and $||K^{n_0}|| < 1$. Consider now the Markov chain with the initial density $\pi(x_0)$, and the transition density $p(x', x)$; N is the random number of the last state. Introduce auxiliary random weights by use of the recursion formula

$$Q_0 = \frac{f(x_0)}{\pi(x_0)}, \qquad Q_n = Q_{n-1}\frac{k(x_{n-1}, x_n)}{p(x_{n-1}, x_n)},$$

and consider the random variable

$$\xi = \sum_{n=0}^{N} Q_n h(x_n).$$

We shall see, under certain assumptions about the functions $\pi(x)$ and $p(x', x)$, that

$$\mathrm{E}\xi = I_h = (\varphi, h).$$

The nature of these assumptions is clear: the paths of the chain must have a nonzero probability of originating at points x with $f(x) \neq 0$, and for the transitions $x' \to x$, if $k(x', x) \neq 0$. That is, we must require

$$\pi(x) \neq 0 \text{ if } f(x) \neq 0, \; p(x', x) \neq 0 \text{ if } k(x', x) \neq 0. \tag{1.7}$$

The random variable ξ is a commonly used and most convenient estimator of the functional $I_h = (\varphi, h)$; we shall therefore call it a basic estimator of (φ, h). Let K_1 be an operator with the kernel function

$$k_1(x', x) = |k(x', x)|.$$

Theorem 1.1. *Under the conditions (1.7) and* $\rho(K_1) < 1, f \in L_1$, $h \in L_\infty$,

$$\mathrm{E}\xi = \mathrm{E}\sum_{n=0}^{N} Q_n h(x_n) = I_h = (\varphi, h).$$

Proof. To average the series termwise, let us reconstruct the Markov chain by introducing a new state coordinate,

$$\delta_n = \begin{cases} 0 & \text{if the transition } x_{n-1} \to x_n \text{ leads to a break of the trajectory,} \\ 1 & \text{otherwise,} \end{cases}$$

so that this chain is formally of infinite length; $\delta_0 = 1$.

$$\text{Let } \Delta_n = \prod_{k=0}^{n} \delta_k = \begin{cases} 1 & \text{until the first break,} \\ 0 & \text{after the first break.} \end{cases}$$

Then we can write

$$\xi = \sum_{n=0}^{\infty} \Delta_n Q_n h(x_n). \tag{1.8}$$

Suppose for the moment that the functions $k(x', x), f(x), h(x)$ are non-negative. Then

$$\mathrm{E}\xi = \sum_{n=0}^{\infty} \mathrm{E}[\Delta_n \, Q_n \, h(x_n)]. \tag{1.9}$$

Now,

$$\begin{aligned}
&\mathrm{E}\ [\Delta_n Q_n h(x_n)] = \mathrm{EE}\,[\Delta_n Q_n h(x_n)|x_0,\dots,x_n)] \\
&= \mathrm{E}[Q_n\varphi(x_n)\mathrm{E}(\Delta_n|x_0,\dots,x_n)] = \mathrm{E}\Big[Q_n h(x_n)\prod_{k=0}^{n-1}(1-p(x_k))\Big] \\
&= \overbrace{\int_X\dots\int_X}^{n+1} h(x_n)\pi(x_0)\Big[\prod_{k=0}^{n-1} r(x_k,x_{k+1})\Big] \\
&\times\Big[\prod_{k=0}^{n-1}\frac{k(x_k,x_{k+1})}{r(x_k,x_{k+1})[1-p(x_k)]}\Big]\frac{f(x_0)}{\pi(x_0)}\Big[\prod_{k=0}^{n-1}[1-p(x_k)]\,\mathrm{d}x_0\dots\mathrm{d}x_n\Big] \\
&= \overbrace{\int_X\dots\int_X}^{n+1} f(x_0)h(x_n)\Big[\prod_{k=0}^{n-1} k(x_k,x_{k+1})\Big]\,\mathrm{d}x_0\dots\mathrm{d}x_n = (K^n f,h),
\end{aligned}$$

because

$$\begin{aligned}
\mathrm{E}[\Delta_n|x_0,\dots,x_n] &= \mathrm{P}(\Delta_n=1|x_0,\dots,x_n) \\
&= \mathrm{P}(\delta_1=\dots=\delta_n=1|x_0,\dots,x_n) \\
&= \prod_{k=0}^{n-1}[1-p(x_k)].
\end{aligned}$$

Thus, for the non-negative functions $k(x',x), h(x), f(x)$, and by use of (1.9), the theorem is proved.

Let us turn to the general case of alternating functions $k(x',x), f(x)$, $h(x)$. Let $Q_n^{(1)}$ be the weights that correspond to (1.1) with $k_1(x',x)=|k(x',x)|, f_1(x)=|f(x)|, h_1(x)=|h(x)|$.

Now

$$|\xi| = \Big|\sum_{n=0}^{\infty}\Delta_n Q_n h(x_n)\Big| \le \sum_{n=0}^{\infty}|\Delta_n Q_n h(x_n)| = \sum_{n=0}^{\infty}\Delta_n Q_n^{(1)} h_1(x_n) = \xi_1.$$

By the previous assumptions, $\mathrm{E}\xi_1=(\varphi_1,h_1)$ is finite. Here, $\varphi_1=K_1\varphi_1+f_1$. Thus, by the Lebesque-dominated convergence theorem, equality (1.9) holds. □

2. Notice that formal substitution of $f(x')=\delta(x'-x), \pi(x')=\delta(x'-x), x_0=x$, into

$$\mathrm{E}\xi = \mathrm{E}\Big[\sum_{n=0}^{N} Q_n h(x_n)\Big] = (\varphi,h) = (f,\varphi^*), \tag{1.10}$$

and putting $Q_0=1$ yields

$$\varphi^*(x) = \mathrm{E}\Big[h(x)+\sum_{n=1}^{N} Q_n h(x_n)\Big] = \mathrm{E}\xi_x \tag{1.11}$$

Thus, we may apply (1.11) to estimate the solution of the conjugate equation at a given point. We can prove equality (1.11) (as the theorem above was proved) by use of the expansion $\varphi^*=\sum_{n=0}^{\infty}K^{*n}h$.

The expression (1.11) is widely used in the Monte Carlo technique to calculate the variances of estimators and to construct estimators for the bilinear functionals of the form $(\varphi, \varphi^* \chi)$ and, in particular, for the functionals of perturbation theory.

It is possible to estimate the solution of equation (1.1) at a given point on the basis of the expression

$$\varphi(x) = f(x) + \mathrm{E} \sum_{n=0}^{N} Q_n^* f(x_n),$$

considering the initial equation as conjugate to $\varphi^* = K^* \varphi^* + h$. But sometimes it is not possible to choose the appropriate Markov chain for the conjugate equation; besides, the last expression allows estimation of $\varphi(x)$ only at one point x. To construct a more effective local estimate it is enough to represent the integral equation as:

$$\varphi(x) = (\varphi, h_x) + f(x),$$

where $h_x(x') = k(x', x)$. Hence,

$$\varphi(x) = \mathrm{E}\Big[\sum_{n=0}^{N} Q_n k(x_n, x)\Big] + f(x).$$

The last expression makes it possible to estimate $\varphi(x)$ at a few points simultaneously.

3. Now let us consider the case of the direct simulation, when $Q_n \equiv 1$. If $f(x) \geq 0$ and $\int_X f(x)\, dx = 1$, $k(x', x) \geq 0$ and $q(x') = \int_X k(x', x)\, dx \leq 1$, then, taking

$$\pi(x) = f(x), \quad p(x', x) \equiv k(x', x),$$

we obtain $Q_n = 1$, $n = 0, 1, \ldots$, and $\xi = \sum_{n=0}^{N} h(x_n)$.

This type of kernel is called a 'substochastic kernel', which occurs when the physical process may be regarded as a Markov chain, for example, a chain of photon collisions in a medium. If additionally $h \equiv 1$, then $\xi = N$ and $\mathrm{E}N < +\infty$, provided $\rho(K) < 1$. Here $K = B_p$, so we can regard the main result of Section 1.2 as a corollary from Theorem 1.

4. Random variable (1.8) is usually called the 'collision estimate'. Another standard estimate for the functional $I_h = (\varphi, h)$ is the random variable

$$\eta = \frac{Q_N h(x_N)}{p(x_N)}, \tag{1.12}$$

which is called the 'absorption estimate'.

Theorem 1.2. *If the conditions of Theorem 1.1 and the relation*

$$p(x) \neq 0, \ x \in \operatorname{supp} h,$$

hold, then $\mathrm{E}\eta = I_h$.

Proof. Clearly,

$$\eta = \sum_{n=0}^{\infty} \Delta_n (1 - \delta_{n+1}) Q_n h(x_n).$$

Using this equality, the proof is constructed exactly as the proof of Theorem 1.1 □

It is interesting that the direct estimate of the average quantity of 'physical captures' is simpler to represent as a collision estimate, after introducing a number of the collision type into the set of phase space X coordinates (see Section 3.1).

5. Let us notice that, usually, the Monte Carlo method is used for solving integral equations (1.1), related to the generalized kernel functions $k(x', x)$. These functions include factors of the 'delta-function' type. This means that the dimension of an integration region is less than the dimension of the phase space X and

$$K \in [L_1^{(1)} \to L_1^{(1)}], \text{ or } K^* \in [L_\infty^{(1)} \to L_\infty^{(1)}],$$

where $L^{(1)}$ is a sufficiently regular functional subspace (for instance, including continuous, or piecewise continuous functions). Here it is necessary to include the same factors in the transition density $p(x', x)$, and to disregard them when constructing the weights Q_n. Theorems 1.1 and 1.2 are clearly extended to this case.

6. All the above and the following statements are extended for solving the system of algebraic equations

$$x = Ax + b,$$

provided $\rho(A) < 1$, by way of the replacements:

$$k(x', x) \to a_{i,j}, \quad r(x', x) \to r_{i,j} = P(j \to i),$$

$$f(x) \to b_i, \ \pi(x) \to \pi_i, \ p(x') \to p_j, \ h(x) \to h_i,$$

$$\varphi = K\varphi + f \to x_i = \sum_{j=1}^{m} a_{i,j} x_j + b_i,$$

$$(\varphi, h) = \int_X \varphi(x) h(x) \mathrm{d}x \to (x, h) = \sum_{j=1}^{m} x_j h_j$$

Here, the Markov chain $i_0, i_1, \ldots, i_N$ related to the finite phase space (1, ..., m) is simulated. It is determined by the transition probabilities $r_{i,j}$, the initial probabilities $\pi_i = P(i_0 = i)$ and the probabilities p_j of terminating in the state j (or at the transition $j \to i$). The collision estimate is equal to

$$\xi = \sum_{n=0}^{N} Q_n h_{i_n}; \quad Q_n = \frac{b_{i_0}}{\pi_{i_0}} \prod_{k=1}^{n} \frac{a_{i_k, i_{k-1}}}{r_{i_k, i_{k-1}}}$$

Under conditions similar to (1.7), we get: $\mathrm{E}\xi = (x, h)$. To evaluate x_i it is enough to use

$$h = (0, \ldots, 0, 1, 0, \ldots, 0),$$

where unity occupies the ith place. For instance, those algorithms are used for solving the finite-difference boundary-value problems (see Section 2.6).

1.4. Variances of the standard estimates

Lemma 1.1. *Under the conditions of Theorem 1.1 for the case of the non-negative functions $f(x), h(x), k(x', x)$ the variance of the collision estimate ξ is given by the expression:*

$$\mathrm{D}\xi = (\chi, h[2\varphi^* - h]) - I^2, \tag{1.13}$$

where χ is the Neumann series for the equation

$$\chi = K_p\chi + p^2/\pi; \tag{1.14}$$

here, K_p is the integral operator with the kernel function $k^2(x', x)/p(x', x)$.

Proof. Firstly, let us notice that the quantities

$$Q_0^2 = \frac{f^2(x_0)/\pi(x_0)}{\pi(x_0)}, \quad Q_n^2 = Q_{n-1}^2 \frac{k^2(x_{n-1}, x_n)/p(x_{n-1}, x_n)}{p(x_{n-1}, x_n)}$$

can be regarded as the standard auxiliary weights related to equation (1.14). We take into account also the equality

$$\mathrm{E}\Big[\sum_{m=n+1}^{\infty} \Delta_m \frac{Q_m}{Q_n} h(x_m) \Big| x_0, \dots, x_n; \Delta_n\Big] = \Delta_n[\varphi^*(x_n) - h(x_n)],$$

which here directly follows from (1.11). On this basis,

$$\begin{aligned}
\mathrm{E}\xi^2 &= \mathrm{E}\sum_{n=0}^{\infty} \Delta_n Q_n^2 h^2(x_n) + 2\mathrm{E}\sum_{n=0}^{\infty}\sum_{m=n+1}^{\infty} \Delta_n\Delta_m \\
&\times Q_n Q_m h(x_n)h(x_m) = (\chi, h^2) \\
&+ 2\sum_{n=0}^{\infty} \mathrm{E}\Delta_n Q_n^2 h(x_n) \mathrm{E}\Big[\sum_{m=n+1}^{\infty} \Delta_m \frac{Q_m}{Q_n} h(x_m) \Big| x_0, \dots, x_n; \Delta_n\Big] \\
&= (\chi, h^2) + 2\mathrm{E}\sum_{n=0}^{\infty} \Delta_n Q_n^2 h(x_n)[\varphi^*(x_n) - h(x_n)] \\
&= (\chi, h^2) + 2(\chi, h[\varphi^* - h]) = (\chi, h[2\varphi^* - h])
\end{aligned}$$

The following statement is related to the case of alternating functions.

Theorem 1.3. *If the conditions of Theorem 1.1 and the relations*

$$\rho(K_p) < 1, \qquad f^2/\pi \in L_1$$

hold, then the finite quantity $\mathrm{D}\xi$ is determined by expression (1.13) and (1.14).

Proof. Clearly, under the Theorem conditions, $\mathrm{D}\xi < +\infty$, where

$$\xi_1 = \sum_{n=0}^{\infty} \Delta_n |Q_n||h(x_n)| \geq |\xi|.$$

Hence, all the operations from the proof of Theorem 1.2 are also allowed here. □

Note that $\rho(K_p) < 1$ implies $\rho(K_1) < 1$ because of the inequality $||K_1^n||^2 \leq ||K_p^n||$.

Theorem 1.4. *Under the conditions of Theorem 1.3, the following relations hold:*

$$\mathrm{E}\xi_x^2 = h(x)[2\varphi^*(x) - h(x)] + \int_X \frac{k^2(x,x')}{p(x,x')} \mathrm{E}\xi_{x'}^2 \mathrm{d}x, \tag{1.15}$$

$$\mathrm{E}\xi^2 = \int_X \frac{f^2(x)\mathrm{E}\xi_x^2}{\pi(x)} \mathrm{d}x. \tag{1.16}$$

Proof. Equation (1.15) is derived exactly as the expression for $\mathrm{E}\xi^2$, and (1.16) is the direct corollary from the total expectation formula. □

The following statements about the variance of an absorption estimate are proved similarly to Lemma 1.1 and Theorem 1.3.

Lemma 1.2. *Under the conditions of Theorem 1.2 for the case of the non-negative functions $f(x), h(x), k(x', x)$ the variance of the absorption estimate is given by the expression*

$$D\eta = (\chi, \frac{h^2}{p}) - I^2, \tag{1.17}$$

where χ is the Neumann series for equation (1.14). □

Theorem 1.5. *If the conditions of Theorem 1.2 and the relations*

$$\rho(K_p) < 1, \; f^2/\pi \in L_1, \; h^2/p \in L_\infty$$

hold, then the finite quantity $\mathrm{D}\eta$ is determined by expressions (1.17) and (1.14). If $x_0 \equiv x$, then $\eta = \eta_x$, and

$$\mathrm{E}\eta_x^2 = h^2(x)/p(x) + \int \frac{k^2(x,x')}{p(x,x')} \mathrm{E}\eta_{x'}^2 \mathrm{d}x', \tag{1.18}$$

$$\mathrm{E}\eta^2 = \int_X \frac{f^2(x)\mathrm{E}\eta_x^2}{\pi(x)} \mathrm{d}x. \tag{1.19}$$

□

Note that relations (1.15), (1.16) and (1.18), (1.19) (which first appeared in [12]) are easily obtained from (1.13), (1.14) and (1.17), (1.14) (which first appeared in [21]) on the basis of the dual representation (1.4) of the linear functionals $(\chi, H), H \in L_\infty$.

The expressions (1.16) and (1.19) for $E\xi^2$ and $\mathrm{E}\eta^2$ show that the uniform (with respect to $x = x_0$) minimization of the functions $\mathrm{E}\xi_x^2$ and $\mathrm{E}\eta_x^2$ also allows solution of the problem of minimizing $\mathrm{D}\xi$ and $\mathrm{D}\eta$ for a function $f(x)$ using the following expressions

$$\pi(x) = c|f(x)|(\mathrm{E}\xi_x^2)^{1/2}, \ \eta(x) = c|f(x)|(\mathrm{E}\eta_x^2)^{1/2}.$$

Finally, note that in the case of the direct simulation when

$$p(x', x) = k(x', x), \quad \pi(x) = f(x), \quad \rho(K) < 1, \quad 0 \le h \in L_\infty,$$

the variance $\mathrm{D}\xi$ is finite because $\chi \equiv \varphi$, and

$$\begin{aligned} \mathrm{E}\xi^2 &= (\varphi, h[2\varphi^* - h]) \le ||\varphi h||_{L_1} \, ||2\varphi^* - h||_{L_\infty} \\ &\le ||\varphi||_{L_1} \, ||h||_{L_\infty} \, ||2\varphi^* - h||_{L_\infty} < +\infty \end{aligned}$$

1.5. The main approaches to variance reduction

Here all the functions are supposed to be non-negative.

1. To construct the collision estimate ξ and the absorption estimate η, one usually makes use of the following transition densities:

$$p_\xi(x', x) = \frac{k(x', x)g(x)}{[K^* g](x')}, \quad p_\eta(x', x) = \frac{k(x', x)g(x)}{g(x')}. \tag{1.20}$$

It is obvious that $\int p_\xi(x', x)\, dx \equiv 1$; therefore, in practical utilization of $p_\xi(x', x)$, it is necessary to additionally introduce slight absorption. Further, if $p_\eta(x', x)$ is used, it is necessary to assume that $K^* g \le g$ because

$$\int_X p_\eta(x', x) dx = 1 - p_\eta(x') = \frac{[K^* g](x')}{g(x')}.$$

Note that $K^* g = g$ can be assumed only at the points where $h(x) = 0$. The initial density is chosen as

$$\pi_g(x) = f(x)g(x)/(f, g). \tag{1.21}$$

For the sake of simplicity, we suppose here that

$$[K^* \varphi^*](x) = \varphi^*(x) - h(x) > 0 \quad \forall x \in X.$$

Theorem 1.6. *If $p(x', x) \equiv p_\xi(x', x)$ and $\pi(x) \equiv \pi_g(x)$ for $g(x) \equiv \varphi^*(x)$, then*

$$\mathrm{E}\xi = I_h, \quad \mathrm{D}\xi = 0.$$

Proof. Clearly, here $\Delta_n \equiv 1$ and $N = +\infty$. Let us denote

$$\eta_0 = Q_0 \varphi^*(x_0), \ \eta_m = \sum_{n=0}^{m-1} Q_n h(x_n) + Q_m \varphi^*(x_m), \ m = 1, 2, \ldots$$

It is easy to show that $\eta_m \equiv I_h$, $m = 0, 1, 2, \ldots$. Indeed,

$$\eta_0 = [f(x_0)/\pi_{\varphi^*}(x_0)]\varphi^*(x_0) \equiv I_h,$$

$$\eta_{m+1} - \eta_m = Q_m h(x_m) + Q_m\Big[\frac{\varphi^*(x_m) - h(x_m)}{\varphi^*(x_{m+1})}\varphi^*(x_{m+1})$$
$$- \varphi^*(x_n)\Big] \equiv 0, \quad m = 0, 1, ...$$

Further,

$$\lim \eta_m = I_h = \xi + \zeta, \quad \zeta = \lim[Q_m \varphi^*(x_m)] \geq 0.$$

Clearly, the latter limit exists. Hence,

$$I_h = \mathrm{E}\xi + \mathrm{E}\zeta = I_h + \mathrm{E}\zeta$$

and $\mathrm{E}\zeta = 0$. Therefore, $\zeta = 0$ and $\xi = I_h$ with probability one. □

Theorem 1.7. *If $p(x', x) \equiv p_\eta(x', x)$ and $\pi(x) \equiv \pi_g(x)$ for $g(x) \equiv \varphi^*(x)$, then*

$$\mathrm{E}\eta = I_h, \quad \mathrm{D}\eta = 0.$$

Proof. Note that here

$$\frac{f(x_0)}{\pi_{\varphi^*}(x_0)} = \frac{I_h}{\varphi^*(x_0)}, \qquad \frac{h(x_N)}{p(x_N)} = \varphi^*(x_N)$$

$$\frac{k(x_{n-1}, x_n)}{p(x_{n-1}, x_n)} = \frac{\varphi^*(x_{n-1})}{\varphi^*(x_n)}, \quad n = 1, 2, ...$$

The direct substitution of these expressions into the equality $\eta = Q_N h(x_n)/p(x_N)$ gives

$$\eta \equiv I_h$$

□

The peculiarity of the Markov chain of equation (1.20) with $g \equiv \varphi^*$ is that

$$r(x', x) = 0 \text{ if } \varphi^*(x) = 0.$$

The last expression does not bias the estimate because, in view of equation (1.11), this estimator scores zero for paths that originate at points $\{x : \varphi^*(x) = 0\}$.

Holton obtained an analogous result using the restriction $(\varphi^* - h)/\varphi^* \leq c < 1$ [30]. This restriction simplifies the proof. Indeed, it is clear that here

$$\xi_m = \sum_{n=0}^{m} Q_n h(x_n) = I_h\Big[1 - \prod_{k=0}^{m} \frac{\varphi^*(x_k) - h(x_k)}{\varphi^*(x_k)}\Big],$$

Hence, by Holton's condition, $\xi = \lim_{m\to\infty} \xi_m = I_h$ with probability one. This method of proof is not applicable to the general case when h may be zero for $\varphi^* \neq 0$ (or when h/φ^* is arbitrarily close to zero). Most Monte Carlo applications in transfer theory are concerned with the general case.

2. We call $\varphi^*(x)$ an importance function (of the point x with respect to the functional I_h), and Monte Carlo techniques similar to the ideal are called importance sampling.

Further, we shall consider some practical uses of the importance sampling principle. First it should be noted that zero-variance importance sampling cannot be used to solve practical problems because, first, the function φ^* is unknown and, second, because $p \equiv 0$, it is necessary to simulate chains of infinite length. Therefore, in practice, an approximate importance sampling based on the change φ^* with $g \approx \varphi^*$ is used [33], [49]. The remarkable peculiarity of this algorithm is its independence of the constant factor of the functions g. Also, absorption is introduced in the mth state, which almost certainly results in a break in the trajectory. If

$$g = \text{const}(1+\varepsilon)\varphi^*, \quad |\varepsilon(x)| \le \delta < 1,$$
$$p(x) = \delta_p, \quad \{p(x) = 0 \text{ if } n \le m\},$$
$$q' = \rho(K) \cdot \frac{1+\delta}{(1-\delta)(1-\delta_p)} < 1,$$

this algorithm gives [49]

$$\mathrm{D}\xi \le c_1\delta^2 + c_2\delta_p\left[\rho(K)\frac{1+\delta}{1-\delta}\right]^m$$

The efficiency of the Monte Carlo methods is measured by $s = 1/(t \cdot \mathrm{D}\xi)$, where t is the average computer time per sample value ξ. It follows from the estimate above that for appropriate m, for example, when $m \sim |\ln\delta|, s \to 0$ if $\delta \to 0$. This conclusion could be used when m is to be chosen. Because information is usually available about a function proportional to φ^* only, the absence of dependence on the constant factor in $g = \text{const}(1+\varepsilon)\varphi^*$ is very useful. The cited relations show that the Monte Carlo techniques under study may be improved by making use of a *priori* information (even though crude) about φ^*. The large m is used for the good approximation of φ^*. The most important applications of this method are those used in problems of transfer theory. In these problems, X is the phase space of coordinates and velocities and the quantity K^*g is in fact not computable for realistic models. However, the kernel function of the integral equation may be represented as the product of the conditional densities of the scattering angle, the free-path length, and other quantities. This makes it possible to use the approximation of the importance function at each step of simulation. It is not difficult to formulate and prove the corresponding zero-variance theorem. This technique often results in simpler formulas than the method of direct utilization of the quantity K^*g. In a similar manner, applications of the asymptotic solution of the Milne problem to the calculation of particle transfer through thick layers of a medium have been developed [49]. It should be noted that there are also other ways of using importance functions, for instance, a direct realization of the importance sampling for estimating the path-integral representation $I_h = \mathrm{E}\xi$. However, the dependence of the importance function on the constant factor, the necessity of calculating the quantity K^*g, and the severe restriction $K^*g/g \le 1$ all create difficulties in the practical realization of these techniques. For example, the unbiased estimate [1]

$$\xi = (f,g) + \sum_{n=0}^{N} Q_n\{h(x_n) + [K^*g](x_n) - g(x_n)\},$$

which is, clearly, analogous to the control-variance method for calculating an integral, seems to be of particular interest, because we may simultaneously use various values of g

when a number of I_h functionals are calculated. But as we have mentioned, it is difficult to use this estimate in complicated calculations.

3. Sometimes it is convenient to optimize only specific transitions, for example, to optimize the free length distribution, when simulating the particle trasfer. The Markov chain, which gives the minimal variance of collision estimates, is determined here by a special nonlinear integral equation [49]. In this way, asymptotic optimization of simulation of the free length is constructed.

The nonlinear theory of optimization gives an interesting result with respect to the total transition $x' \to x$: if the absorption is simulated with the probability $p(x') \neq 0$, then the optimal modification of the transition $x' \to x$ is determined [49] by equation (1.3) with the kernel function:

$$k_1(x', x) = k(x', x)[1 - p(x')]^{-1/2}.$$

The total optimization of the collision estimate for problems with alternating function is also determined by a special nonlinear integral equation [49], [8].

4. Now, we consider an optimization of weight estimates of the solution $\varphi(x)$ using the following functions $\{h_i\}$:

$$h_i(x) = \begin{cases} 1/mes(D_i), & x \in D_i \\ 0, & x \notin \mathrm{D}_i \end{cases}$$

for partition $\{D_i\}_{i=1,\dots,n}$ of the domain X. The expressions for $\mathrm{E}\eta_i^2$ and $\mathrm{E}\xi_i^2$ directly imply that

$$\mathrm{E}\eta_i^2 = O(\Delta_n), \quad \mathrm{E}\xi_i^2 = O(\Delta_n^{-1}) \;\text{ for }\; \chi(x) < c < +\infty$$

$$\Delta_n \equiv \max_i \; mes \; D_i$$

as $\Delta_n \to 0$, while

$$\mathrm{E}\eta_i = \mathrm{E}\xi_i \approx \varphi(x_i) = O(1)$$

where x is a middle point of the subdomain D_i, provided $\varphi(x)$ is continuous in $D_i, i = 1, \dots, n$. Therefore, it is sufficient to minimize the quantities $\mathrm{E}\eta_i^2$ and $\mathrm{E}\xi_i^2$, as $\Delta_n \to 0$. Besides, under sufficiently general assumptions of the kernel $k(x, x')$ [for example, when $k(x, x')$ is continuous with respect to x' in $\mathrm{D}_i, i = 1, \dots, n$] we have

$$h_i(2\varphi_i^* - h_i) \sim h_i^2, \quad \Delta_n \to 0.$$

Therefore, when absorption estimates are used it is reasonable to choose transition densities such that

$$\inf \sup_x \{\chi(x)/p(x)\}$$

is attained. Likewise, $\inf \sup_x \chi(x)$ must be attained in the case of collision estimates. It is easy to prove the corresponding limit statements, taking into account that in the optimal variants $\chi/p \equiv$ const and $\chi =$ const (see the text below). According to this, it is expedient to choose $\{D_i\}$ so that $\varphi_i^* \approx h_i$ with sufficient accuracy.

Further, we consider a more general problem of uniform optimization of the estimate for $\rho(x)\varphi(x)$, assuming the functions $\rho\varphi$ and $\rho^2\chi$ to satisfy the conditions formulated above for φ and χ, respectively. On the other hand, we construct a class of transition densities assuming that, for class $\{p_\eta\}$, equation (1.14) takes the form:

$$\chi(x) = \int\limits_X \frac{k(x',x)g(x')}{g(x)}\chi(x')\mathrm{d}x' + \frac{f(x)}{g(x)}(f,g). \tag{1.22}$$

We thus formulate the following problem of uniform optimization of absorption and collision estimates.

A. In the case of absorption estimates, when constructing the transition density of the type p_η, it is necessary to find a function g such that

$$\inf_g \sup_x \{\rho^2(x)\chi(x)/p(x)\} \tag{1.23}$$

is attained.

B. In the case of collision estimates, when using the transition density of the type p_η, it is necessary to find g such that

$$\inf_g \sup_x \{\rho^2(x)\chi(x)\} \tag{1.24}$$

is attained.

Denote the class of measurable non-negative functions of X by F. In what follows, we use

Lemma 1.3. *Let $\psi \in F, f_i \in F, i = 1, ..., n$. Then*

$$\inf_{\varphi\in F} \sup_{i,x} \{f_i(x)(\varphi,\psi)/\varphi(x)\} = (\psi, \max_i f_i)$$

it attained for

$$\varphi(x) = \varphi^*(x) = \max_i f_i(x), \quad x \in X.$$

Proof. This lemma can be formulated as a problem of minimizing the quantity (φ, ψ) in the class of functions $F_0 \subset F$ satisfying the condition

$$\sup_x \{\max_i f_i(x)/\varphi(x)\} = 1.$$

On account of

$$(\varphi, \psi) \ge \left(\max_i f_i, \psi\right) \quad \forall \varphi \in F_0$$

Lemma 1.3 is proved. □

In the special case where $n = 1$, Lemma 1.3 gives

$$f(x)(\varphi^*, \psi)/\varphi^*(x) = (f, \psi).$$

Multiplying equation (1.22) by $g(x)$ yields

$$\chi(x) = \frac{\varphi(x)}{g(x)}(f,g). \tag{1.25}$$

Since $p = (g - K^*g)/g$, we have

$$\frac{\rho^2(x)\chi(x)}{p(x)} = \frac{\rho^2(x)\varphi(x)(f,g)}{g(x)-[K^*g](x)} = \frac{\rho^2(x)\varphi(x)(\varphi,[I-K^*]g)}{g(x)-[K^*g](x)}. \tag{1.26}$$

The function $g - K^*g$ can be considered as an arbitrary function from $F \cap L_\infty(X)$. Hence, using equation (1.26) and Lemma 1.3 we obtain

Proposition 1. *If $\rho^2\varphi \in F\cap L_\infty(X)$, then for $p(x',x) \equiv p_\eta(x',x)$, the quantity of equation (1.23) is attained if $g = \varphi^* = K^*\varphi^* + \rho^2\varphi$. In this case, $\rho^2\chi/p \equiv (\varphi, \rho^2\varphi)$.*

For the collision estimate, the function χ is also determined by equation (1.25)(if $p \equiv p_\eta$), and the following proposition holds:

Proposition 2. *If $K^*(\rho^2\varphi) \le \rho^2\varphi$, then for $p(x',x) \equiv p_\eta(x',x)$, the quantity of equation (1.24) is attained for $g \equiv \rho^2\varphi$; in this case, $\rho^2\chi \equiv (f, \rho^2\varphi)$.*

The comparison of Propositions 1 and 2 shows that from the standpoint of the uniform estimation of the solution in the case of a large mean number of collisions S_0, the optimal collision estimate is essentially more effective than the optimal absorption estimate because

$$(\varphi,\rho^2\varphi)/(f,\rho^2\varphi) \approx S_0.$$

It is obvious that a specific value of a small absorption probability does not essentially affect the cost of the uniform algorithms. Therefore, we can consider the collision estimate as a main (from the practical point of view) Monte Carlo estimator if

$$p(x',x) = (1-p_0)\frac{k(x',x)\varphi^*(x)}{[K^*\varphi^*](x')}$$

$$\pi(x) = f(x)\varphi^*(x)/(f,\varphi^*),$$

where $0 < p_0 \ll 1$, and $\varphi^* = \rho^2\varphi$ if $K^*(\rho^2\varphi) \le \rho^2\varphi$. Otherwise, it is reasonable to take $\varphi^* = K^*\varphi^* + \rho^2\varphi$, since the corresponding absorption collision is a minimax one; recall that the collision estimate must be S_0 times more effective than the absorption estimate. Analysis of the effectiveness of the utilization of the asymptotic solutions in the statistical modelling of radiative transfer processes [49] shows that it is reasonable to apply the technique described. Note that the inequality $K^*\varphi \le \varphi$ is definitely valid if $K^* = K$. This condition is also satisfied in a model problem of transfer theory we consider in Section 1.11.

In conclusion, let us remark that it is possible during real computing to estimate the minimax values of $\rho^2\chi$ and $\rho^2\chi/p$, because $(f,\rho^2\varphi)$ is the standard linear functional (see Section 1.3) and $(\varphi,\rho^2\varphi)$ is the standard bilinear functional (see [48], p.55). So we can approximately compare the algorithm really used with the ideal minimax one using the expressions:

$$(f,\rho^2\varphi) = \mathrm{E}\sum_{n=0}^{N} Q_n f(x_n)\rho^2(x_n), \quad (\varphi,\rho^2\varphi) = \mathrm{E}\sum_{n=0}^{N} Q_n^* f(x_n)\sum_{m=0}^{n}\frac{\rho^2(x_m)Q_m}{Q_m^*}$$
$$= \mathrm{E}\sum_{n=0}^{N} Q_n\rho^2(x_n)\sum_{m=n}^{N}\frac{Q_m^*}{Q_n^*}f(x_m),$$

where $\{Q_n^*\}$ are related to the conjugate equation (see Section 1.3). The latter expression can easily be derived by using double averaging.

The above constructions may be useful for optimizing the global estimation of the solution φ. In particular, the relation

$$||\varphi - \mathrm{E}\tilde{\varphi}||_{L_2} \leq C_1 t^4 + C_2 \frac{\max \chi}{nt^k}$$

holds (see [50], Section 1.5), where $\tilde{\varphi}$ is the linear prolongation of the histogram $\{\varphi_i\}$, n is the sample size and t is the grid step.

1.6. The use of recurrent representations

1. Let us consider the integral equation

$$\varphi(x) = \int_X k(x, x')\varphi(x')\mathrm{d}x' + h(x) \tag{1.27}$$

or in the operator form

$$\varphi = K\varphi + h,$$

where $X \in R_k$ is a compact set, $h \in L_\infty(X)$, $K \in [L_\infty \to L_\infty]$.

In this section, we assume that the spectral radius $\rho(K) < 1$. In so doing, the solution to equation (1.27) is expressed as the Neumann series, but here we construct and investigate the basic recurrent estimate of the Monte Carlo methods by means of a direct probabilistic approximation of equation (1.27). We show below that this approach is especially effective with branching Markov chains, where the complete expression of the estimate is of necessity cumbersome. In addition, the recurrent representations of estimates and second-order moments allow one to use the Bellman principle for their optimization [50].

Here the appropriate parametrization of the initial problem is of great importance. For equation (1.27), that means utilizing the equation

$$\varphi_\lambda = \lambda K \varphi_\lambda + h, \quad 0 < \lambda < 1/\rho(K). \tag{1.28}$$

Clearly, the solution to the equation $\varphi_\lambda \equiv \varphi(x, \lambda)$ depends on λ monotonically and even strictly monotonically except for trivial degenerated cases.

To construct the estimate using the Monte Carlo method we introduce the terminating Markov chain $\{x_n\}$, $n = 0, 1, ...N$, with the transition density $p(x, x')$, where the quantity

$$p(x) = 1 - \int_X p(x, x')\mathrm{d}x' \geq 0$$

is considered as the probability of the trajectory terminating in the transition from x to x'; N is a random number of the last state of the chain. The Markov chain is related to equation (1.27) by the weight factor $q(x, x')$. For the transition from x to x', this factor is calculated by the formula

$$q(x, x') = \frac{k(x, x')}{p(x, x')} \tag{1.29}$$

and $q(x, x') = 0$ if this transition terminates the trajectory. To simplify the mathematics, we can formally assume that after termination,the trajectory gets into a special absorbing point and remains there. Relation (1.29) implies, in particular, that the following known condition (see Section 1.3) should be met:

$$p(x, x') \neq 0 \quad \text{for} \quad k(x, x') \neq 0. \tag{1.30}$$

We have to construct an unbiased random estimate of the solution for the chain originating at the point x, i.e. to construct such a functional ξ that $E\xi = \varphi(x)$. Obviously, this estimate allows the following randomization of equation (1.27):

$$\xi_0 = h_0 + q(x_0, x_1)\xi_1, \tag{1.31}$$

where $h_0 = h(x_0)$ and $\xi_i = \xi$ for $x = x_i$; one can easily see that

$$\begin{aligned} E[q(x_0, x_1)\xi_1] &= EE[q(x_0, x_1)\xi_1 | x_1] \\ &= E[q(x_0, x_1)\psi(x_1)] = [K\psi](x_0), \quad \psi \equiv E\xi, \end{aligned}$$

i.e. relation (1.31) is a probabilistic approximation of equation (1.27) in the sense that, in average, (1.31) coincides with (1.27). However, recurrent estimate (1.31) is unbiased only if the additional condition is met:

$$E\xi < C < +\infty \quad \forall x \in X, \tag{1.32}$$

because the solution to equation (1.27) is unique in L_∞. It is interesting to note that here we have an assertion similar to the known Lax theorem: from the probabilistic approximation and stability of (1.32) follows the convergence (in the sense of validity of the law of large numbers). In order to deduce the unbiasedness of ξ directly from (1.31), we note that when continuing recursion (1.31) up to the end of the trajectory $\{x_n\}$, we obtain the known representation of the collision estimate ξ (see Section 1.3). For parametrized equation (1.28), the representation is of the form:

$$\xi_0(\lambda) = \sum_{n=0}^{\infty} \lambda^n Q_n h(x_n), \tag{1.33}$$

where

$$Q_0 \equiv 1, \quad Q_n = Q_{n-1} q(x_{n-1}, x_n).$$

We will point out rather apparent properties of the function $E\xi(\lambda)$. As a function of λ (at any fixed $x = x_0$), it is continuous up to $\psi = \infty$, since this function is the expectation, i.e. integral of monotonous with respect to λ and non-negative function of the trajectory over the probabilistic measure. The integral $E\xi(\lambda)$ allows differentiation under the sign E, because as a result one obtains an integral of a non-negative monotone function. Therefore, $E\xi(\lambda)$ is a convex function of λ and (see, for example, [65]) $\|\xi(\lambda)\|$ is a convex function of λ. Further, since X is a compact set, the same property is valid for the supremum of $\|\xi(\lambda)\|$ from the continuity of $E\xi(\lambda)$ at $+\infty$ for any x. That means that there exists such a sequence $\{\lambda_k\}$, that

$$\lim_{k\to\infty} \|\xi(\lambda_k)\| = +\infty, \quad \|\xi(\lambda_k)\| < +\infty .$$

Therefore, there exist such values λ_k that the value of $\|\xi(\lambda_k)\|$ is as large as one likes but finite, and $\|\xi(\lambda)\|$ is continuous if $\lambda < \lambda_k$ [65] as it is a convex function.

The experience of solving various problems by means of the Monte Carlo method shows that one can always construct such a function $p(x, x')$ that the following useful condition is met:

$$|q(x, x')| < C < +\infty . \tag{1.34}$$

Theorem 1.8. *If $\rho(K) < 1$, $k(x, x') \geq 0$, $h(x) \geq 0$ and conditions (1.30), (1.34) are met, then recursion (1.31) defines the unbiased estimate ξ.*

Proof. The function $\|E\xi(\lambda)\|$ is monotonic and continuous with respect to λ up to $+\infty$ and by virtue of condition (1.34) we have $\|E\xi(\lambda)\| < +\infty$ for sufficiently small $\lambda > 0$. Therefore, if $\|E\xi\| = +\infty$, then one can find such $\lambda < 1$ that $\|\varphi\| < \|E\xi(\lambda)\| < +\infty$.

However, in so doing, we have $E\xi(\lambda) = \varphi_\lambda$; but $\|\varphi_\lambda\| \leq \|\varphi\|$ due to the monotonic dependence of φ_λ on λ. The contradiction obtained proves Theorem 1.8. □

Theorem 1.8 is a well-known statement, but it is of interest because in its proof one uses only the recurrent nature of the estimate and does not perform cumbersome calculations that involve detailing the Neumann series and preliminary averaging of the complete expression for ξ over 'absorbing-surviving' (see Section 1.3).

Theorem 1.8 can obviously be extended to the case with alternating signs, provided the condition $\rho(K_1) < 1$ is met, where K_1 is an integral operator with the kernel $|k(x, x')|$, since in this case we have

$$|\xi_0| \leq \xi_0^{(1)} = |h_0| + |q(x_0, x_1)|\xi_1^{(1)}$$

which means that $|E\xi| < C < +\infty$ and direct averaging of relation (1.31) gives the unbiasedness of ξ.

2. The variance $D\xi = E\xi^2 - \varphi^2$ is determined by the value $E\xi^2$ that depends significantly on the choice of the transition density $p(x, x')$. Let us consider the relation

$$\xi_0^2 = h_0^2 + 2h_0 q(x_0, x_1)\xi_1 + q^2(x_0, x_1)\xi_1^2 . \tag{1.35}$$

We can obtain from (1.35) a simpler recursion if we replace the term $2h_0 q(x_0, x_1)\xi_1$ by the corresponding expectation

$$E[2h_0 q(x_0, x_1)\xi_1] = 2h_0[K\varphi]_0 = 2h_0(\varphi_0 - h_0).$$

Let us show that such substitution does not affect the value of the mean square. Considering (1.35) as a recursion for ξ^2 and continuing it to the end of a trajectory, we obtain the expression

$$\xi_0^2 = \sum_{n=0}^{\infty} Q_n^2[h_n^2 + 2h_n q(x_n, x_{n+1})\xi_{n+1}].$$

When averaging this expression one can (see Section 1.3) replace the bracketed term by its conditional expectation at a fixed value of x_n, i.e. by the quantity $h_n(2\varphi_n - h_n)$.

Thus, we have $E\xi^2 = E\zeta^2$ and

$$\zeta_0^2 = h_0(2\varphi_0 - h_0) + q^2(x_0, x_1)\zeta_1^2 \,. \tag{1.36}$$

Therefore, if the conditions of Theorem 1.8 are satisfied and $p(K_p) < 1$ (K_p is an integral operator with the kernel $k^2(x,x')/p(x,x')$), then the function $\psi \equiv \mathrm{E}\xi^2$ is determined by the equation

$$\psi = h(2\varphi - h) + K_p\psi \tag{1.37}$$

which was first obtained in [12] when constructing Monte Carlo methods for solving systems of algebraic equations. One can easily show (see Section 1.4) that $\|K_1\| < \|K_p\|$; hence under the mentioned conditions (1.37) is also valid in the case with alternating signs. In the case of constant signs, the most general result is as follows: the Neumann series for equation (1.37) determines the quantity $\mathrm{E}\xi^2$ for the unbiased estimate ξ whether or not the condition $p(K_p) < 1$ is met (see Section 1.4).

Note that for solving some applied problems, for example, problems in radiation transfer theory, one has estimate linear functionals of the form:

$$(f, \varphi) = (\varphi^*, h) = \int_X \varphi^*(x)h(x)\mathrm{d}x,$$

where $f \in L_1$ and φ^* is the solution to the conjugated equation $\varphi^* = K^*\varphi^* + f$; this equation is considered as the basic one in radiation transfer theory.

Suppose the probability density $\pi(x)$ is such that $\pi(x) \neq 0$ for $f(x)\varphi(x) \neq 0$ and the point x distributed with the density $\pi(x)$. Then

$$(f, \varphi) = \mathrm{E}\zeta, \quad \zeta = \frac{f(x_0)}{\pi(x_0)}\xi_0.$$

One can easily see that the expression for ζ coincides with the standard collision estimate

$$\zeta = \sum_{m=0}^{N} Q_n h(x_n), \quad Q_0 = \frac{f(x_0)}{\pi(x_0)}$$

for the quantity (φ^*, h) (see Section 1.3). According to the Bellman principle, minimizing the variance of this estimate is performed in the following way: one determines the optimal variant of the transition density $p(x,x')$, as was shown above, then one determines the corresponding function $\psi \equiv \mathrm{E}\xi^2$; the optimal density $\pi(x)$ is then determined by the formula:

$$\pi(x) = \frac{|f(x)|\psi^{1/2}(x)}{\int f(x)\psi^{1/2}(x)\mathrm{d}x} \,.$$

3. When one constructs the basic unbiased estimate ξ, the use of branching Markov chains in solving integral equations is worthwhile in some cases. First and foremost this is true for integral equations describing the particle transfer with multiplication. In addition, trajectory 'splitting' can be used to decrease the variance and to make easier the calculation of the collision estimate. To solve such problems, an expression for the variance ξ^2 with 'splitting' (i.e. artificial branching) was constructed in [7] by averaging the corresponding recurrent representation. Considerations in the present Section show that such a technique needs additional justification, due to the fact that the condition $\|\mathrm{E}\xi^2\| < +\infty$ must be met. Here we give an expression for $\mathrm{E}\xi^2$ in a somewhat different form, which is more convenient in modelling physical processes of particle transfer with multiplication.

Let us represent the kernel of integral equation (1.27) and the transition density $p(x,x')$ as follows:

$$k(x,x') = \sum_{i=0}^{m} p_i(x) k_i(x,x')\nu_i(x'), \quad p(x,x') = \sum_{i=0}^{m} p_i(x) p_i(x,x'),$$

$$p_i(x) \geq 0, \quad \sum_{i=0}^{m} p_i(x) \equiv 1,$$

where $p_i(x,x')$ are the transition densities. Here $\nu_i(x')$ is an integer-valued function that can also have zero value, which means that a trajectory terminates at the point x'. Modelling the transition with branching is performed in the following way. A random number i of the transition probability $p_i(x,x')$ is chosen according to the probabilities $\{p_i(x)\}$, the transition $x \to x'$ is modelled, the weight is multiplied by the quantity $q_i(x,x') = k_i(x,x')/p_i(x,x')$ and then $\nu_i(x')$ independent trajctories with the obtained initial weight are constructed from the point x'. The corresponding recurrent representation for ξ is of the form:

$$\xi_0 = h_0 + q_i(x,x') \sum_{k=1}^{\nu_i(x')} \xi_1^{(k)}.$$

Here the justification of unbiasedness (i.e. the relation $\mathrm{E}\xi \equiv \varphi$) presented in this Section is especially appropriate, since the full expression and direct averaging of the estimate ξ are more cumbersome than in the case without branching. By averaging the recurrent representation of ξ^2 we obtain the following expression for the function $\psi = \mathrm{E}\xi^2$, in the same manner as above,

$$\psi(x) = h(x)[2\varphi(x) - h(x)] + \int \sum_{i=1}^{m} p_i(x) \frac{k_i^2(x,x')\nu_i(x')[\nu_i(x') - 1]}{p_i(x,x')} \varphi^2(x')\mathrm{d}x'$$

$$+ \int \sum_{i=1}^{m} p_i(x) \frac{k_i^2(x,x')\nu_i(x')}{p_i(x,x')} \varphi(x')\mathrm{d}x.$$

As above, apart from the condition $\rho(K_1) < 1$, we need here the condition $\rho(K_p) < 1$, where the kernel of the operator K_p is

$$\sum_{i=1}^{m} \frac{k_i^2(x,x')\nu_i(x')}{p_i(x,x')} . \tag{1.38}$$

It can be seen from expression (1.38) that for the problem with non-negative functional elements, the operator K_p coincides with K when $p_i(x,x') \equiv k_i(x,x')$. Therefore, if $\rho(K) < 1$, then the variance of direct modelling (i.e. estimates with $Q_n \equiv 1$) is finite. Note that for a physical process of particle transfer with branching, the relation $\rho(K) < 1$ implies that the system is undercritical. In particular, for an undercritical system, it is finite if one considers $h(x)$ as the probability of particle absorption at a point x.

4. Let us consider a nonlinear integral equation of the form:

$$\varphi(x) = \int_X k(x,x')\varphi^n(x')\mathrm{d}x' + h(x) \tag{1.39}$$

or in the operator form:

$$\varphi = K\varphi^n + h. \tag{1.40}$$

We consider this equation in $L_\infty(X)$. Let us also consider the parametrized equation

$$\varphi_\lambda = \lambda K \varphi_\lambda^n + h, \quad \lambda \geq 0. \tag{1.41}$$

First, we assume all the terms in the problem to be non-negative. We suppose that the following condition is met:

$$q_n = n\lambda \|K\| < 1 \tag{1.42}$$

and for $\lambda = \lambda_0$ there exists such a solution φ_{λ_0} that $\|\varphi_{\lambda_0}\| < 1$. In so doing, the solution to equation (1.41) that satisfies the additional condition $\|\varphi_\lambda\| < 1$ is unique for $\lambda \leq \lambda_0$. Indeed, let us assume that there are two such solutions $\varphi_\lambda^{(1)}$ and $\varphi_\lambda^{(2)}$.

Then the following relation is valid:

$$\varphi_\lambda^{(1)} - \varphi_\lambda^{(2)} = \lambda K \Big[(\varphi_\lambda^{(1)} - \varphi_\lambda^{(2)}) \sum_{k=0}^{n-1} (\varphi_\lambda^{(1)})^k (\varphi_\lambda^{(2)})^{(n-1-k)} \Big] = K_\varphi [\varphi_\lambda^{(1)} - \varphi_\lambda^{(2)}].$$

It is obvious that with the constraint (1.42) we have $\|K_\varphi\| < 1$ and therefore $\varphi_\lambda^{(1)} \equiv \varphi_\lambda^{(2)}$.

One can easily show that the solution φ_λ exists and monotonically increases with λ for $\lambda \leq \lambda_0$. Really, if $\lambda_1 \leq \lambda_2 \leq \lambda_0$, then

$$\lambda_1 K \varphi_{\lambda_2}^n + h \leq \varphi_{\lambda_2}.$$

Thus, iterations of equation (1.41) beginning with $\varphi_0 \equiv \varphi_{\lambda_2}$ monotonically decrease and converge to the solution for $\lambda = \lambda_1$ and $\varphi_{\lambda_1} \leq \varphi_{\lambda_2}$.

Note that the Monte Carlo algorithms for solving (1.40) were first constructed in a more general form in [18] on the basis of the connection between integral equations with power nonlinearity and branching Markov chains. These chains are constructed in the same manner as those in Section 1.3 (i.e. for $n = 1$), with the difference that after the next transition $x \to x'$, n independent trajectories are constructed from the point x', with the weight previously multiplied by a weight factor, which we assume here to be bounded:

$$q(x, x') = \frac{k(x, x')}{p(x, x')} \leq c < +\infty. \tag{1.43}$$

The basic estimate can be defined by the following recurrent formula

$$\xi_0 = h_0 + q(x_0, x_1) \prod_{k=1}^{n} \xi_1^{(k)}, \tag{1.44}$$

where $\{\xi_1^{(k)}\}$ are independent (in the aggregate) values of the estimate that correspond to the above-mentioned trajectories originating at the point x'. According to the above conclusions, condition (1.42) provides the termination of a branching trajectory with the probability 1 (and even the finiteness of the average number of branches that is necessary in order to make calculations possible) and thereby provides the possibility to continue recursion (1.44) up to the end of the trajectory. In addition, for small values of λ, the value of $\|E\xi\|$ is small as well, since the following relation is valid, provided that the inequality $\lambda n < 1$ is true:

$$\xi_\lambda \leq \frac{\|h\|}{1 - \lambda n}.$$

Theorem 1.9. *If $k(x, x') \geq 0$, $h(x) \geq 0$, $\lambda_0 \geq 1$ and conditions (1.30), (1.42), (1.43) are met, then recursion (1.44) defines an estimate ξ such that*

$$\mathrm{E}\xi \equiv \varphi, \quad \|\varphi\| < 1.$$

Proof. As with $n = 1$, the direct averaging of relation (1.44) for $n > 1$ gives equation (1.40). Therefore, if the inequality $\|\mathrm{E}\xi\| < 1$ is valid, the conclusion of the theorem is true. Let us now suppose that $\|\mathrm{E}\xi\| \geq 1$. The properties of the function $\|\mathrm{E}\xi(\lambda)\|$ mentioned above for $n = 1$ are valid for $n > 1$ as well. Therefore, one can find such a $\lambda < 1$ that the following inequality is valid:

$$\|\varphi\| < \|\mathrm{E}\xi(\lambda)\| \leq 1.$$

In this case, $\mathrm{E}\xi(\lambda) \equiv \varphi_\lambda$ and $\|\varphi_\lambda\| \leq \|\varphi\|$, since $\lambda < 1$. The contradiction obtained proves the theorem. □

The unbiasedness of the obtained estimate is proved in [18] with the more general suggestion of the convergence of iterations for equation (1.40), i.e. the estimate of an iterative solution is considered. In so doing one uses a cumbersome representation of the estimate as a whole and the Neumann series for equation (1.40) (see [19]). Note that the estimate ξ can be realized numerically by using, for instance, the well-known lexicographic scheme (see, for example, [19]).

The direct averaging (as with $n = 1$) of recurrent representation of ξ^2 in (1.44) gives the following equation (see also [69]) for $\psi \equiv \mathrm{E}\xi^2$:

$$\psi = h(2\varphi - h) + K_p\psi^n. \tag{1.45}$$

Here, K_p is an integral operator with the kernel $k^2(x, x')/p(x, x')$. For the function $\mathrm{E}\xi^2$ one can obtain a statement similar to Theorem 1.9, if one additionally demands that condition (1.42) be met with the substitution $K \to K_p$, $\varphi \to \psi$ and takes equation (1.45) instead of (1.37).

The transition to the case with alternating signs is performed in the following way as with $n = 1$ (see Section 1.4). Initially, the reasoning is made for the problem with the kernel $k(x, x')$ and the free term $h(x)$. Thus, one obtains a majorant for ξ and uses it in averaging recursion (1.44). As a result, one obtains the function $\mathrm{E}\xi$ that satisfies equation (1.40) and inequality $\|\mathrm{E}\xi\| < 1$, i.e. the solution sought for. Arguments for $\mathrm{E}\xi^2$ are similar.

1.7. Randomization

1. Various examples introducing additional randomness for constructing effective simulation algorithms can be found in the literature devoted to Monte Carlo methods (see, for example, [19]). This section is concerned with randomized algorithms for estimating probabilistic characteristics of equations with random parameters.

Randomized estimation for the statistical moments of the solution is presented below. Assume functional equation $L\phi = f$ to be solved by the Monte Carlo method on the basis of simulation of a stochastic process (denote the trajectories of this process by ω). This means that random variables $\xi_k(\omega)$ are constructed so that

$$\mathrm{M}\xi_k(\omega) = J_k, \quad k = 1, 2, ..., m,$$

where J_k are the functionals of ϕ to be evaluated (M denotes the mathematical expectation).

Let the operator L and the function f depend on a random field σ (for example, a random medium in transfer theory, random force in elasticity theory, etc.). Also,

$$\xi_k = \xi_k(\omega,\sigma), \quad J_k = J_k(\sigma)$$

and

$$\mathrm{M}[\xi_k(\omega,\sigma)|\sigma] = J_k(\sigma),$$

where the variables ω and σ are generally not independent.

Consider the problem of evaluating the quantities

$$J_k = \mathrm{E}J_k(\sigma), \quad R_{kj} = \mathrm{E}[J_k(\sigma)J_j(\sigma)], \quad k,j = 1,...,m,$$

where E denotes the mathematical expectation with respect to the distribution of σ.

The following obvious method is known for evaluating these mathematical expectations. First, realizations of σ are constructed: then the equation $L_\sigma\phi_\sigma = f_\sigma$ is solved precisely enough for each realization by a numerical or an analytical technique. Finally, statistical estimates of the desired quantities are calculated. However, this approach fails for complicated multi-dimensional problems because the computational cost of an explicit solution of the equation considered is too high. Therefore, it is useful to apply sometimes a method of 'double randomization'. In our case, this technique follows from the following relations:

$$\begin{aligned} \mathrm{E}J_k(\sigma) &= \mathrm{EM}\xi_k(\omega,\sigma) = \mathrm{M}_{(\omega,\sigma)}\xi_k(\omega,\sigma), \\ \mathrm{E}[J_k(\sigma)J_j(\sigma)] &= \mathrm{M}_{(\omega_1,\omega_2,\sigma)}[\xi_k(\omega_1,\sigma)\xi_j(\omega_2,\sigma)], \end{aligned} \tag{1.46}$$

where ω_1 and ω_2 are conditionally independent trajectories, constructed for one fixed realization of σ and the subscript of the expectation symbol indicates the distribution to which it corresponds. Clearly, we have to assume the existence of the total expectations exposed in (1.46), i.e.

$$M_{(\omega,\sigma)}|\xi_k(\omega,\sigma)| < +\infty, \quad M_{(\omega_1,\omega_2,\sigma)}[|\xi_k(\omega_1,\sigma)\xi_j(\omega_2,\sigma)|] < +\infty.$$

Relations of equation (1.46) show that to estimate the quantities J_k it is sufficient to construct only one trajectory for a fixed σ, while the estimation of the quantities R_{kj} requires two conditionally independent trajectories. To optimize the randomization technique, it is natural to use the 'splitting method' (see, for example, [49]). In this method, the quantities J_k are estimated as follows. First, one constructs n conditionally independent trajectories (i.e. a vector $\omega = (\omega_1,...,\omega_n)$, with σ fixed), and then a random variable

$$\zeta_k^{(n)}(\omega,\sigma) = \frac{1}{n}\sum_{i=1}^{n}\xi_k(\omega_i,\sigma)$$

is used instead of $\xi_k(\omega,\sigma)$. The optimal value of n is calculated by the formula (see, for example, [49])

$$n = \sqrt{\frac{a_2}{a_1}\cdot\frac{t_1}{t_2}},$$

where

$$a_1 = \mathrm{E}(\mathrm{M}\xi_k)^2 - J_k^2, \quad a_2 = \mathrm{ED}\xi_k\ ,$$

t_1 is the average computing time for a fixed realization of σ, and t_2 is the average computing time for a fixed realization of ω.
2. As a particular case of using the formulae (1.46), it is possible to consider randomization of the collision estimate (see Sect. 1.3)

$$\xi = \sum_{n=0}^{N} Q_n h(x_n),$$

where

$$Q_0 = \frac{f(x_0)}{\pi(x_0)}, \quad Q_n = Q_{n-1}\frac{k(x_{n-1}, x_n)}{p(x_{n-1}, x_n)}$$

Let $\tilde{k}(x_{n-1}, x_n), \tilde{f}_0, \tilde{h}_n$ be independent unbiased estimates of the corresponding values $k(x_{n-1}, x_n), f(x_0), h(x_n)$ (for instance, random estimates of integrals, which express those values), $\tilde{Q}_n$ is the corresponding unbiased estimate of the weight Q_n and K_1 is the integral operator with the kernel function $\mathrm{E}|\tilde{k}(x', x)|$. If $\rho(K_1) < 1, \mathrm{E}|\tilde{h}| \in L_\infty, \mathrm{E}|\tilde{f}| \in L_1$, then [19]

$$I_h = (\varphi, h) = \mathrm{M}\tilde{\xi},$$

where $\tilde{\xi} = \sum\limits_{n=0}^{N} \tilde{Q}_n \tilde{h}_n$. Besides, in [19] it is shown that

$$\mathrm{E}\tilde{\xi}^2 = (\chi, h[2\varphi^* - h]) + (\chi, \mathrm{D}\tilde{h}),$$

where χ is the Neumann series for the equation

$$\chi'(x) = \int\limits_X \frac{\mathrm{E}\tilde{k}^2(x', x)}{p(x', x)}\chi'(x')dx' + \frac{\mathrm{E}\tilde{f}^2(x)}{\pi(x)}, \quad \text{or} \quad \chi' = K_p'\chi' + \mathrm{E}(\tilde{f}^2/\pi),$$

if $\rho(K_p') < 1, \mathrm{E}\tilde{f}^2/\pi \in L_1$.

If evaluating $h(x)$ is only randomized, then

$$\mathrm{D}\tilde{\xi} = \mathrm{D}\xi + (\chi, \mathrm{D}\tilde{h}),$$

where χ is determined in the same way as in Section 1.4. As well as in Section 1.4 it is possible to show that the relation $\rho(K_p') < 1$ implies the relation $\rho(K_1') < 1$.
3. Further, a class of Monte Carlo algorithms for solving large-scale linear algebraic systems with dense matrices based on randomization of matrix-vector multiplication is considered [9]. Varying the number of non-zero rows in random sparse matrices involved from N to 1 (N being the number of equations) one can proceed from the deterministic successive approximations method to the Neumann-Ulam scheme [19] (with transition probabilities p_{ij} all equal to $1/N$); thus, the statistical error being easily controlled. The general scheme with sufficient conditions for variance boundness is considered. The method is intended for solving stochastic problems.

Consider a system of linear algebraic equations

$$Au = g; \tag{1.47}$$

here $u, g \in \mathbb{R}^N$, A is a nonsingular $N \times N$ matrix. Suppose that the stationary iterative method for solving this system can be constructed. It means that (1.47) can be transformed to

$$u = Ku + f,$$

with the spectral radius of K less than unity, and thus the successive approximations

$$u^{(n+1)} = Ku^{(n)} + f$$

$$u^{(0)} \quad \text{given}$$

converge to the unique solution of (1.47). The results below are independent of whether the elements of matrix K are available in explicit form or not.

Let $S^{(1)}, S^{(2)}, \ldots, S^{(n)}, \ldots$ be an infinite sequence of independent realizations of the random matrix S such that $\mathrm{E}S = K$. Define the sequence of random vectors $\xi^{(n)}$ setting

$$\begin{aligned} \xi^{(n+1)} &= S^{(n)}\xi^{(n)} + f, \\ \xi^{(0)} &= u^{(0)}, \end{aligned} \tag{1.48}$$

Since $S^{(n)}$ and $\xi^{(n)}$ are independent within this construction, we get $\mathrm{E}\xi^{(n)} = u^{(n)}$ for all n. For covariance matrix

$$X^{(n)} = \mathrm{E}\Big(\xi^{(n)} - u^{(n)}\Big)\Big(\xi^{(n)} - u^{(n)}\Big)^T,$$

the following equation can be obtained:

$$X^{(n+1)} = \mathrm{E}\Big(SX^{(n)}S^T\Big) + C^{(n)}; \tag{1.49}$$

here, $C^{(n)}$ is defined by

$$C^{(n)} = \mathrm{E}\Big((S-K)u^{(n)}(u^{(n)})^T(S-K)^T\Big).$$

The general theory of matrix equations then gives us the following theorem:

Theorem 1.10. *The sequence $X^{(n)}$ of covariance matrixes is bounded if the Kronecker product of the random matrix S with itself satisfies the condition*

$$\rho(\mathrm{E}(S \otimes S)) < 1. \tag{1.50}$$

□

Under weak additional assumptions, this condition is not only sufficient but also necessary. Another inequality

$$\rho(\mathrm{E}(S^T S)) < 1 \tag{1.51}$$

can be shown to be a sufficient condition for boundness of the sequence $\mathrm{E}\|\xi^{(n)} - u^{(n)}\|_2$, and hence, the boundness of the sequence $X^{(n)}$ too.

Let $J=\{j_1, j_2, \ldots, j_L\}$ be a random set of L different natural numbers less than or equal to N, where j_1 is chosen with equal probabilities among all these numbers, j_2 is chosen with equal probabilities among the numbers remaining, etc. For all first indexes i put

$$s_{ij} = \begin{cases} \frac{N}{L} k_{ij}, & \text{if } j \in J, \\ 0 & \text{otherwise.} \end{cases}$$

Thus, the random matrix S constructed has L non-zero rows which are equal to corresponding rows of matrix K, multiplied by N/L. For all i, j

$$\mathrm{E}s_{ij} = \frac{N}{L} k_{ij} \cdot P\{j \in J\} = k_{ij}.$$

Due to the special form of random matrix S, to calculate random vector $\xi^{(n+1)}$ (see (1.48)) one needs only L components of $\xi^{(n)}$ to be calculated on the previous step with, in turn, only L components of $\xi^{(n-1)}$ involved. Hence, the amount of computational work needed to calculate k components of $\xi^{(n+1)}$ is proportional to $kL + nL^2$. If $L = 1$, then the algorithm described coincides with the standard Neumann-Ulam scheme (or collision estimate) with transition probabilities all equal to $1/N$. If $L = N$, it becomes the successive approximation method with no randomization at all (vectors $\xi^{(n)}$ are equal to $u^{(n)}$ for all n).

For the random matrix S considered above, equation (1.49) can be rewritten in the form

$$X^{(n+1)} = \frac{N}{L}\frac{L-1}{N-1} K X^{(n)} K^T + \frac{N}{L}\frac{N-L}{N-1} K D^{(n)} K^T + C^{(n)}, \tag{1.52}$$

where $D^{(n)}$ is the diagonal of $X^{(n)}$. For the elements $c_{ij}^{(n)}$ of $C^{(n)}$ in this case, the equation

$$c_{ij}^{(n)} = \frac{1}{L}\frac{N-L}{N-1}\left(N \sum_{l=1}^{N} k_{il} k_{jl} (u_l^{(n)})^2 - \sum_{l,m=1}^{N} k_{il} u_l^{(n)} u_m^{(n)} k_{jm}\right)$$

holds.

Thus, the matrixes $C^{(n)}$ are decreased a little faster than $1/L$ and are all equal to zero for $L = N$.

To analyze the dependence of $X^{(n)}$ of L, consider two linear operators G_1 and G_2 defined for an arbitrary $N \times N$ matrix X by

$$G_1 X = K X K^T, \; G_2 X = N K D K^T,$$

where D is the diagonal of X, and put

$$G = \frac{N}{L}\frac{L-1}{N-1} G_1 + \frac{1}{L}\frac{N-L}{N-1} G_2. \tag{1.53}$$

With this notation, $G = \mathrm{E}(S \otimes S)$ and (1.52) takes the simple form

$$X^{(n+1)} = G X^{(n)} + C^{(n)}$$

or

$$X^{(n+1)} = G^n C^{(0)} + G^{n-1} C^{(1)} + \ldots + C^{(n)}. \tag{1.54}$$

It can be shown that for the random sparse matrix S, the inequality

$$L > N \| K^T \|_2^2$$

is sufficient for (1.50) to hold. The similar condition

$$L > N\|K\|_2^2$$

is sufficient for (1.51) to be true in this case. Note that for symmetric matrix K

$$\|K^T\|_2 = \|K\|_2 = \rho(K) .$$

From (1.54) it follows that if $\rho(G) \leq 1$ (that is, L is sufficiently large) then $X^{(n)}$ decrease a little faster than $1/L$ due to the behaviour of $C^{(k)}, k = 1, ..., N$ as a function of L. If $\rho(G) > 1$ for small L then the second term in (1.53) dominates, because $\rho(G_1) = \rho(K \otimes K) < 1$ and the factor $\frac{N}{L}\frac{L-1}{N-1}$ does not exceed unity. The factor $\frac{1}{L}\frac{N-L}{N-1}$ at the G_2 decreases a little faster than $1/L$. Therefore, $X^{(n)}$ decreases much faster than $1/L$ for $n \geq 2$. Moreover, if $\rho(G) \gg 1$, then $X^{(n)}$ decreases as $\frac{1}{L^n}$.

Thus, one can expect that L is optimal if $\rho(G) \sim 1$.

1.8. Vector estimates related to the triangular system of integral equations

1. Vector estimates for the case of a triangular matrix kernel are considered below. Consider in L_∞ a system of the form

$$\varphi_i = h_i + \sum_{j=1}^{i} K_{ij}\varphi_j, \quad i = 1, ..., m, \tag{1.55}$$

where K_{ij} is an integral operator with the kernel $k_{ij}(x, y)$:

$$[K_{ij}\varphi](x) = \int_X k_{ij}(x, y)\varphi(y)dy.$$

In the operator form we rewrite system (1.55) as $\Phi = H + \mathbf{K}\Phi$. Here

$$\|\Phi\|_{L_\infty} = \sum_{i=1}^{m} \int_X |\varphi_i(x)|dx.$$

Theorem 1.11. *If K_{ij} are bounded operators, then the inequality*

$$\rho(\mathbf{K}) \leq \max_i \rho(K_{ii}) = \rho_0$$

holds. Furthermore, if $k_{ij}(x, y) \geq 0$, then $\rho(\mathbf{K}) = \rho_0$

Proof. Let us consider the system

$$\varphi_i = h_i + \lambda \sum_{j=1}^{i} K_{ij}\varphi_j, \quad i = 1, ..., m,$$

This system is uniquely solvable if $0 < \lambda < \rho_0^{-1}$. Therefore, $\rho(\mathbf{K}) < \rho_0$ (see, for example, [13]). If $k_{ij}(x, y) \geq 0$, then $\sup \|K_{ii}^n\| < \|\mathbf{K}^n\|$ and $\rho_0 \leq \rho(\mathbf{K})$. This completes the proof. □

Note that the first statement of Theorem 1.11 holds for arbitrary linear operators.
2. Let us consider the possibility of constructing and analyzing vector estimators in the Monte Carlo method on the basis of the relations obtained above. For constructing vector estimators of the solution of the system of integral equations $\Phi = \mathbf{K}\Phi + H$ we define a terminating Markov chain $x = x_0, x_1, \ldots, x_N$ with transition probability density $p(x, y)$ different from zero on the support of the matrix kernel $K(x, y)$, that is, on the union of the supports of the kernels $k_{ij}(x, y)$.

We introduce also the stochastic matrix weight:

$$\tilde{Q}_0 = \{\delta_{ij}\}, \quad \tilde{Q}_n = \tilde{Q}_{n-1} K(x_{n-1}, x_n)/p(x_{n-1}, x_n), \quad n = 1, 2, \ldots,$$

where δ_{ij} is the Kronecker symbol. Under the additional condition $\rho(\mathbf{K}_1) < 1$ (where $\mathbf{K}_1$ is the operator obtained from $\mathbf{K}$ by replacing the kernels k_{ij} by their moduli) we have

$$\mathrm{E}\xi_x = \Phi(x), \qquad \xi_x = \sum_{n=0}^{N} \tilde{Q}_n H(x_n).$$

For the covariance matrix $\Psi(x) = \mathrm{E}[\xi_x, \xi_x']$ the matrix-integral equation

$$\Psi(x) = [H\Psi' + \Psi H' - HH']_x + \int \frac{K(x, y)\Psi(y)K'(x, y)}{p(x, y)} \mathrm{d}y, \tag{1.56}$$

or $\Psi = \mu + \mathbf{K}_p\Psi$ has been obtained in [49] under the hypothesis that the Neumann series for the equation $\Psi = \mu_1 + \mathbf{K}_{p,1}\Psi$ converges, where the matrix-integral operator $\mathbf{K}_{p,1}$ and the function μ_1 correspond to the problem with the moduli of the functional characteristics.

Let us denote by $K_{p,ij}$ an integral operator with the kernel $k_{ij}^2(x, y)/p(x, y)$.

Theorem 1.12. *If $K_{p,ii}$ are bounded operators, then for system (1.55) the following inequality holds*

$$\rho(\mathbf{K}_p) \le \max_i \rho(K_{p,ii}) = \rho_0^{(p)}.$$

If, in addition, $k_{ij} \ge 0$, then $\rho(\mathbf{K}_p) = \rho_0^{(p)}$.

Proof. By the special enumeration of elements of the matrix $\Psi(x)$, it is possible to expose (1.56) as a triangular system with the operators $K_{p,ii}$ on the diagonal. In addition, the inequality

$$||K_{p,ij}||^2 \le ||K_{p,ii}|| \, ||K_{p,jj}||$$

holds. □

The Monte Carlo method is ordinarily used to estimate linear functionals of the form

$$I = (F, \Phi) = \int F'(x)\Phi(x)\mathrm{d}x,$$

where $F'(x) = (f_1(x), \ldots, f_m(x))$, with

$$||F||_{L_1} = \sum_j \int |f_j(x)|\mathrm{d}x < +\infty .$$

Let the point x_0 be distributed with probability density $\pi(x)$ such that $\pi(x) \ne 0$ if $F'(x)\Phi(x) \ne 0$. Then, obviously,

$$I = \mathrm{E}\left[\frac{F'(x_0)}{\pi(x_0)}\xi_{x_0}\right] = \mathrm{E}\sum_{n=0}^{N}\frac{F'(x_0)}{\pi(x_0)}Q_n H(x_n) = \mathrm{E}\sum_{n=0}^{N} H'(x_n)Q'_n\frac{F(x_0)}{\pi(x_0)}.$$

The random vector with weight

$$Q_n^{(')} = Q'_n F(x_0)/\pi(x_0)$$

is computed by the formula

$$Q_n^{(')} = [K'(x_{n-1}, x_n)/p(x_{n-1}, x_n)]Q_{n-1}^{(')}.$$

Setting $\zeta = F'(x_0)\xi_{x_0}/\pi(x_0)$, we have

$$\mathrm{E}\zeta^2 = \mathrm{E}[F'(x_0)\xi_{x_0}\xi'_{x_0}(x_0)F(x_0)/\pi^2(x_0)]\,.$$

Thus the variance of ζ is determined by the matrix-valued function $\Psi(x) = \mathrm{E}[\xi_x\xi'_x]$, which we have studied above. Because of the relation

$$\mathrm{E}[F'(x_0)\Psi(x_0)F(x_0)/\pi^2(x_0)] \leq ||\Psi||_{L_\infty}\int_X \Big[\Big(\sum_{i=1}^{m}|f_i(x)|\Big)^2/\pi(x)\Big]\,\mathrm{d}x,$$

for the value $\mathrm{E}\xi^2$ to be finite it is necessary, in addition to the condition of Theorem 1.12, to assume that the condition

$$\int_X \Big[\Big(\sum_{i=1}^{m}|f_i(x)|\Big)^2/\pi(x)\Big]\,dx < +\infty \tag{1.57}$$

holds.

1.9. Calculation of parametric derivatives and the main eigenvalues of integral operators

1. Consider in $L_\infty(X)$ an integral equation of the second kind with a kernel depending on a parameter λ:

$$\varphi(x,\lambda) = \int_X k(x,y,\lambda)\varphi(y,\lambda)\mathrm{d}y + h(x,\lambda). \tag{1.58}$$

Rewrite equation (1.58) in the operator form

$$\varphi = K\varphi + h.$$

Although the integral operator in equation (1.58) has conjugate form, it is denoted here by the symbol K for simplicity.

In equation (1.58), X is an n-dimensional Euclidean space. Consider the problem of calculating the derivatives

$$\varphi^{(n)}(x,\lambda) = \frac{\partial^n(x,\lambda)}{\partial\lambda^n}.$$

Introduce the following notation: $K^{(n)}$ is the integral operator with the kernel $k^{(n)}(x,y,\lambda)$, and $\rho(K)$ is the spectral radius of the operator K. It is assumed that the functions $k^{(n)}$ and $h^{(n)}$ are measurable with respect to x. By formally differentiating the relation in equation (1.58) n-times with respect to λ, we obtain a triangular system of integral equations

$$\varphi^{(n)} = \sum_{i=0}^{n} C_n^i K^{(n-i)} \varphi^{(i)} + h^{(n)}, \quad n = 0, 1, ..., m \tag{1.59}$$

or in the operator form, $\Phi = \mathbf{K}\Phi + H$.

Now we give a justification of the utilization of equation (1.59) for calculating the derivatives $\varphi^{(n)}$.

Introduce the additional notation: $F_m(f)$ is the column vector consisting of the derivatives of the function $f(\lambda)$ with respect to λ, i.e.

$$F_m(f) = (f, f^{(1)}, ..., f^{(m)})^T,$$

$D_m(f) = d_{ni}$ is the lower triangular $(m+1)$-dimensional matrix with entries

$$d_{ni} = C_n^{n-i} f^{(n-i)}, \quad i = 0, 1, ..., n; \quad n = 0, ..., m.$$

Using the Leibnitz formula for the mth derivative of the product of two functions, we obtain

Lemma 1.4. *The following relation holds:*

$$F_m\left(\Pi_{i=0}^k f_i\right) = D_m(f_k) \cdots D_m(f_1) F_m(f_0).$$

□

Theorem 1.13. *Suppose that the inequalities*

$$|k^{(n)}(x, y; \lambda')| \le k_n(x, y), \quad |h^{(n)}(x, \lambda)| \le h_0(x), \quad h_0 \in L_\infty \tag{1.60}$$

hold in the interval $\lambda - \varepsilon \le \lambda' \le \lambda + \varepsilon$ for some $\varepsilon > 0$, the integral operators K_n with the kernels $k_n(x, y)$ are bounded, $n = 0, ..., m$ and $\rho(K_0) < 1$. Then $\rho(\mathrm{K}) < 1$ and the functions $\varphi^{(n)}$ satisfy system equation (1.59), $n = 0, 1, ..., m$.

Proof. Following from the Theorem 1.11, the relation $\rho(\mathbf{K}) \le \rho(K_0) = \rho_0 < 1$ holds in the interval $(\lambda - \varepsilon, \lambda + \varepsilon)$. By Lemma 1.4, the corresponding Neumann series for system equation (1.59) coincides with the series obtained by the formal differentiation of the Neumann series for equation (1.58). As follows from equation (1.60), this series is uniformly majorized in $(\lambda - \varepsilon, \lambda + \varepsilon)$. The Neumann series for the majorant is finite, thus proving Theorem 1.13. □

Lemma 1.4 and Theorem 1.13 can be directly extended to the case of vector parameter $\lambda = (\lambda_1, ..., \lambda_k)$, since the corresponding systems of type equation (1.59) can also be written in a triangular form with the diagonal entries coinciding with K. To do this, it is sufficient to choose an indexation of the functions $\varphi^{(i_1,...,i_k)}$ according to the natural order in which they are written in the equations. This kind of indexation is universal provided that the equations are written in accordance with the natural increasing order of the derivatives (note that inside a fixed order of the derivative, the indexation can be arbitrary).

Consider now the Monte Carlo algorithms for solving the constructed triangular systems of integral equations. Theorem 1.12 shows that the variances of vector algorithms are bounded if the spectral radius of the operator with the kernel $k^2(x, y, \lambda)/p(x, y)$ is less than unity ($n = 1, ..., m$), i.e. if the standard variance finiteness condition for the scalar Monte Carlo estimates for equation (1.58) is satisfied.

Let $\xi_x(\lambda)$ be the standard collision estimate (see Section 1.3) for equation (1.7).

Theorem 1.14. *Under the conditions of Theorem 1.13 and relation (1.7) the following statements hold:*
1) $\mathrm{E}\xi_x^{(m)}(\lambda) = \mathrm{E}\sum_{n=0}^{N}[Q_n(\lambda)h(x_n, \lambda)]^m = \varphi^{(m)}(x, \lambda)$;
2) if all the functions in equation (1.58) and their necessary derivatives are non-negative, then it is possible to use $\varepsilon = 0$ *in the conditions of Theorem 1.13;*
3) if $\rho(K_p) < 1$, *then* $\mathrm{D}\xi^{(m)} < +\infty$, $m = 0, 1, 2, ...$

Proof. According to Lemma 1.4, a series, obtained by the differentiation of the estimate $\xi_x(\lambda)$, numerically coincides with the standard vector estimate for system (1.59). So, statements (1) and (3) follow from Theorems (1.13) and (1.12). Under the additional condition of statement (2), the replacement of differentiation and averaging is certainly possible, and statement (1) holds for $\varepsilon = 0$. □

Theorem 1.14 is clearly extended to the functionals (see Section 1.13):

$$(f, \varphi_\lambda) = \mathrm{E}[(f(x_0)/\pi(x_0))\xi_{x_0}(\lambda)], \quad \varphi_\lambda \equiv \varphi(x, \lambda).$$

In particular, if $\rho(K_p) < 1$, and $f^2/\pi \in L_1$, then $\mathrm{D}[(f(x_0)/\pi(x_0))\xi_{x_0}^{(m)}(x)] < +\infty$. The equality

$$\mathrm{E}[(f(x_0)/\pi(x_0))\xi_{x_0}(\lambda)] = (f, \mathrm{E}\xi^{(m)}(\lambda)) = (f, \varphi_\lambda^{(m)})$$

is a consequence of the Fubini theorem because $f \in L_1, \varphi^{(m)} \in L_\infty$, and the equality

$$(f, \varphi_\lambda)^{(m)} = (f, \varphi_\lambda^{(m)})$$

holds because of the relation $||\varphi_{\lambda'}^{(m)}|| < c < +\infty$ for $\lambda' \in [\lambda - \varepsilon, \lambda + \varepsilon]$. Under the condition of statement (2) of Theorem 1.14, the value $\varepsilon = 0$ is admitted here.

2. Let us now mention some problems where this technique of calculating derivatives can be applied. Suppose that σ is a random vector and the functionals $J_k(\sigma)$ which we study depend on $\sigma_i (i = 1, ..., r)$ almost linearly. Note that the last assumption can be considered in the neighbourhood of the point $\sigma^{(0)} = M\sigma$ (the probability of hitting this neighbourhood is close to unity). Then the correlation moments of the vector $J = J_k(\sigma)$ and the autocorrelation moments of the vectors σ and J can be calculated by the linearization of $J(\sigma)$ using the Monte Carlo evaluations of the derivatives $\partial J_k/\partial\sigma_i$. The calculation of the higher order derivatives can thus be used here to control the numerical results and to make them more accurate. In [41] we present also examples of iterative numerical solution of an inverse problem of finding σ by linearizing the equation

$$J_k(\sigma) = J_k^{(0)}, \quad k = 1, ..., n \geq r,$$

where $J_k^{(0)}$ are known (measured) values of functionals.

Now let us apply the technique under consideration to the estimation of the iterates of the resolvent. To this end, in $L_\infty(X)$, consider equation (1.58) of the form:

$$\varphi(x,\lambda) = \lambda^{-1}\int_X k(x,y)\varphi(x,\lambda)dy + \lambda^{-1}h(x)$$

for $|\lambda| > \rho(K)$. We have

$$\varphi_\lambda = [\lambda - K]^{-1}h$$

and

$$\varphi_\lambda^{(m)} = \frac{d^m\varphi_\lambda}{d\lambda^m} = (-1)^m m![\lambda - K]^{-(m+1)}h.$$

On the other hand,

$$\varphi_\lambda^{(m)} = \mathrm{E}\zeta_x^{(m)}, \quad \zeta_x^{(m)} = \sum_{n=0}^{N}(-1)^m\frac{(n+m)!}{n!}\lambda^{-(n+1+m)}Q_nh(x_n).$$

The unbiasedness of the estimate $\zeta_x^{(m)}$ and the relationship $\mathrm{D}\zeta_x^{(m)} < +\infty$ are evidently ensured by Theorem 1.14. Finally, we get

$$[\lambda - K]^{-(m+1)}h = (-1)^m\varphi_\lambda^{(m)}/m! = \mathrm{E}\xi_x^{(m)}$$

$$\xi_x^{(m)} = \sum_{n=0}^{N} C_{n+m}^m Q_nh(x_n)\lambda^{-(n+1+m)}.$$

Under rather general conditions (see, for example, [13]) we have

$$\frac{([\lambda - K]^{-(m+1)}h, f)}{([\lambda - K]^{-m}h, f)} \to \frac{1}{\lambda - \lambda^*},$$

where λ^* is the eigenvalue of the operator K, closest to λ. If $f(x)$ is the probability density, then

$$([\lambda - K]^{-(m+1)}h, f) = \mathrm{E}\xi_x^{(m)}$$

and x is a random point distributed according to the density f.

Note that the corresponding Monte Carlo algorithm for estimating the eigenvalues has been suggested in [45], where $1/\lambda$ was considered instead of λ and the question of whether the variance is finite has not been discussed.

Now, consider the problem of estimating the main eigenvalue of the operator K_p which is related to the kernel $k^2(x,y)/p(x,y)$ and which determines the values $\mathrm{E}\xi_x^2$ (see Section 1.4). Clearly, all the above relations may be used to solve this problem after substituting $Q_n^2h^2(x_n)$ for $Q_nh(x_n)$. Here the additional condition $\rho(K_{p,2}) < 1$, where $K_{p,2}$ is an operator with the kernel $\lambda^{-4}k^4(\cdot,\cdot)/p^3(\cdot,\cdot)$, is required.

To estimate the main eigenvalue of the matrix-integral operator $\mathbf{K}$ (see Section 1.8) one can use the relation

$$(F', [\lambda - \mathbf{K}]^{-(m-1)}H) = \sum_{n=0}^{N} C_{n+m}^m\lambda^{-(n+1+m)}R_n,$$

where $R_n = F'(x_0)\tilde{Q}_nH(x_n)/\pi(x_0)$. It is interesting that the same relation, after the substitution of R_n^2 for R_n, can be used in the case of the operator $\mathbf{K}_p$ (see Section 1.8) because of the equality

$$\mathrm{E}\sum_{n=0}^{N} C_{n+m}^{m} \lambda^{-(n+1+m)} R_n^2 = \Big(\frac{F'}{\pi}, [\lambda - \mathbf{K}_p]^{-(m-1)}(HH')F\Big),$$

since

$$R_n^2 = \frac{F'(x_0)}{\pi(x_0)} \tilde{Q}_n H(x_n) H'(x_n) \tilde{Q}'_n \frac{f(x_0)}{\pi(x_0)}$$

1.10. Test integral equations and problems

1. Let us consider the following mathematical model of the particle transfer with anisotropic scattering. Particles travel from the point $x = 0$ along the x-axis in random steps distributed with the density function $e^{-x}, x > 0$. At the end point of a step, a collision takes place after which 'scattering' (with the probability $1-q$) may occur. Here 'scattering' results in further travel directly forward, i.e. the 'δ - scattering' is supposed. At the point $x = H$, an escape occurs, i.e. the trajectory terminates. It is necessary to evaluate the probability P of particle escape.

The transition density for the collision chain is determined here by the expression

$$k(x', x) = q\exp(-(x - x')), \quad x' \leq x \leq H.$$

The initial collision density is equal to:

$$f(x) = \exp(-x), \qquad 0 \leq x \leq H.$$

Hence, the total collision density $\varphi(x)$ satisfies the equation [19], [41]

$$\varphi(x) = q\int_0^x e^{-(x-x')}\varphi(x')\mathrm{d}x' + e^{-x}. \tag{1.61}$$

Clearly, here $||K|| < q$, and $P = \varphi(H)$ [41]. So, it is nesessary to evaluate the solution of equation (1.61) at one point only. In particular, for this purpose it is possible to use the 'local estimate', which in the case of the direct simulation takes the form (see Section 1.3):

$$\xi = \sum_{n=0}^{N} qe^{-(H-x_n)}, \qquad \mathrm{E}\xi = P - e^{-H}.$$

By direct substitution in (1.61), it is possible to verify the expressions:

$$\varphi(x) = e^{-(1-q)x}, \quad P = e^{-(1-q)H}, \quad 0 \leq x \leq H.$$

The solution of the conjugate equation

$$\varphi^*(x) = q\int_x^H e^{-(x'-x)}\varphi^*(x')\mathrm{d}x' + qe^{-(H-x)}$$

is equal to

$$\varphi^*(x) = e^{-(1-q)(H-x)}.$$

Further,

$$\mathrm{E}\xi^2 = (\varphi, h[2\varphi^* - h]) = q^2\Big\{\frac{1+2q}{1+q}e^{-(1-q)H} - 2e^{-(2-q)H} + \frac{e^{-2H}}{1+q}\Big\}$$
$$= q^2\Big\{\frac{1+2q}{1+q}P - 2e^{-H}P + \frac{e^{-2H}}{1+q}\Big\}.$$

For the simplest 'physical' estimate η, related to the probabilities

$$P(\eta = 1) = P, \quad P(\eta = 0) = 1 - P,$$

we have $\mathrm{E}\eta^2 = P$, and

$$\mathrm{E}(\eta - e^{-H})^2 = P - 2e^{-H}P + e^{-2H}.$$

So, for large values of H, the simplest estimate is better than the local one, if it is necessary to estimate the probability P. But, for the global estimation of the solution, especially with respect to the metric C, the local estimate can have an advantage (see [50], Section 1.5).

2. The collision density in the case of an infinite homogeneous layer $x \geq 0$ with a 'δ-scattering' and a survival probability q, for a source with the density $f(x) = e^{-x}$ satisfies the equation

$$\varphi(x) = q\int_0^x e^{-(x-x')}\varphi(x')\mathrm{d}x' + e^{-x}.$$

The adjoint operator K^* has the form

$$[K^*\varphi](x) = q\int_x^\infty e^{-(x'-x)}\varphi(x')\mathrm{d}x'. \tag{1.62}$$

Next we present the minimax values of the functions χ/p and χ for $p(x', x) = p_\eta(x', x)$ (see Section 1.5).

(1) *Absorption estimate.* The solution of the equation (see Section 1.5)

$$\varphi^* = K^*\varphi^* + \varphi, \qquad \varphi(x) = \exp\{-(1-q)x\},$$

is given by

$$\varphi^*(x) = \frac{2-q}{2(1-q)}\exp\{-(1-q)x\}$$

for which

$$\frac{\chi}{p} \equiv (f, \varphi^*) = (\varphi, \varphi) = \frac{1}{2(1-q)}.$$

(2) *Collision estimate for $p = p_\eta$.* By equation (1.62), $K^*\varphi \leq \varphi$; therefore, we can use $p_\eta(x', x)$ with $g \equiv \varphi$. In this case,

$$\chi = (\varphi, f) = \frac{1}{2-q} \leq \frac{1}{2(1-q)}.$$

3. Let us consider some analogue of the kinetic Boltzmann equation:

$$\varphi(x) = q\int_0^x \varphi(x')\varphi(x')\exp\Big\{-\int_{x'}^x \varphi(s)\mathrm{d}s\Big\}\mathrm{d}x' + \exp\Big\{-\int_0^x \varphi(s)\mathrm{d}s\Big\} \tag{1.63}$$

$$q \in [0,1].$$

This equation can be treated in terms of the transfer theory in the following way: a particle moves from the point $x = 0$ along the x-axis by random runs distributed with the density $\varphi(x)\exp(-\int_0^x \varphi(s)ds)$, $x > 0$. At the end of the path, a collision occurs. As a result, the particle can be absorbed with the probability $1-q$; in the opposite case it moves further. At the point $x = 1$ the particle escapes, i.e. the path terminates. The function $\varphi(x)$ can be treated as the collision density. By the direct substitution of the function into equation (1.63) we are able to verify that

$$\varphi(x) = \exp\Big\{-(1-q)\int_0^x \varphi(s)ds\Big\} = \frac{1}{(1-q)x+1}.$$

The choice of equation (1.63) as a model one is explained by the existence of the exact solution of the equation itself and a set of linear integral equations used in the iteration methods [61]. In addition, the process of modelling the particle path for equation (1.63) constitutes part of the process of modelling the paths for solving the Boltzmann equation. This allows one to consider this equation as a test problem to check the properties of the proposed algorithms for solving the Boltzmann equation.

4. Now, we consider a nonhomogeneous, spherically symmetric, pure-scattering medium illuminated with parallel radiation flux. The problem is to calculate the integral I_0 of the radiance over all directions, at the center of the sphere. The medium is bounded by a sphere S of radius R. Let the initial radiance be $(\pi R^2)^{-1}$, i.e. there is a unit power source on the surface of the sphere. The density of the surface source is equal to (see [41]) the radiance of the incident flux multiplied by the cosine of the angle between the photon direction and the inner normal to the surface at the point of photon incidence. Consequently, in our case the initial flux is equal to

$$\Phi_0(\mathbf{r},\omega) = \frac{1}{\pi R^2}\Delta_S(\mathbf{r})\delta(\omega-\omega_0)\frac{|\mathbf{r}\cdot\omega|}{R},$$

where $\Delta_S(\mathbf{r})$ is a generalized function that corresponds to integration over the sphere S. The functional to be found is

$$I_0 = \int_\Omega \Phi(0,\omega)\,d\omega = (\Phi, p),$$

where $p(\mathbf{r},\omega) = \delta(\mathbf{r})$. Hence, by the theorem of optical mutuality [16, 9],

$$I_0 = (\Phi^*, \Phi_0) = \frac{1}{\pi R^2}\int_S \Phi^*(\mathbf{r}(s),\omega_0)\frac{|\mathbf{r}(s)\cdot\omega_0|}{R}\,ds.$$

Here, Φ^* is the solution of the transfer equation [41], [50] for an isotropic source of density $p(\mathbf{r},\omega) = \delta(\mathbf{r})$. For this source, the function

$$Q(\omega_0) = \int_S \Phi^*(\mathbf{r}(s),\omega_0)\frac{\mathbf{r}(s)\cdot\omega_0}{R}\,ds,$$

is the integral angular density of photons that escape from the medium. In this case, $Q(\omega_0) \equiv Q = \text{const}$ and

$$\int_{\Omega} Q \mathrm{d}\omega = 4\pi Q = \int_{R}\int_{\Omega} p(\mathbf{r},\omega)\,\mathrm{d}\mathbf{r}\,\mathrm{d}\omega = 4\pi; \quad \text{hence} \quad Q \equiv 1.$$

Finally, we have $I_0 = (\pi R^2)^{-1}$. This result is obvious in the case when the scattering coefficient is zero. It remains true for an arbitrary, pure-scattering medium. For the net flux $I_s(0)$ of photons having at least one collision,

$$I_s(0) = \frac{1 - e^{-\tau}}{\pi R^2},$$

where τ is the optical length of the radius of the sphere S. The theorem of optical mutuality is useful when the detector is fixed and the source is extended in phase space.

1.11. The extension of unbiasedness conditions

1. Consider in $L_\infty(X)$ the integral equation

$$\varphi(x) = \int_X k(x,x')\varphi(x')\mathrm{d}x' + h(x), \tag{1.64}$$

or, in the operator form, the equation $\varphi = K\varphi + h$, where $X \subset R^n$; the operator K is assumed to be bounded with the operator K_1 whose kernel is equal to $|k(x,x')|$. It is also assumed that the quantity $\varphi(x)$ can be represented by the Neumann series $\varphi(x) = \sum_{n=0}^{\infty} K^n h(x)$ for a given point x.

To construct the algorithm for the Monte Carlo method, the terminating Markov chain $\{x_n\}$ with the density $p_n(x_n, x')$ of the transition from x_n to x_{n+1} is introduced, and the quantity

$$p_n(x_n) = 1 - \int_X p_n(x_n, x')\,\mathrm{d}x' \geq 0$$

is considered as the probability of a break of the trajectory in the state $x_n, n = 0, 1, ...N$, and N is the random number of the last state of the chain. Further, the weights $\{Q_n\}$

$$Q_0 = 1,$$

$$Q_n = Q_{n-1}\frac{k(x_{n-1}, x_n)}{p_{n-1}(x_{n-1}, x_n)} = Q_{n-1}q_{n-1}(x_{n-1}, x_n)$$

and the standard collision estimate

$$\xi_x = \sum_{n=0}^{N} Q_n h(x_n), \qquad x_0 \equiv x, \tag{1.65}$$

are introduced. It is additionally assumed that

$$p_n(x_n, x') \neq 0 \quad \text{if} \quad k(x_n, x') \neq 0, \tag{1.66}$$

and the weight factors $\{q_n(x_n, x')\}$ are Lebesgue measurable in $X \times X$. This fact (with the boundedness of the operator K_1) guarantees the integrability and the greater measurability of ξ_x with respect to the corresponding probability measure for a fixed value of N; hence, ξ_x is measurable, and the quantity $\mathrm{E}|\xi_x|$ makes sense.

To justify the unbiasedness of ξ_x, i.e. the relation $\mathrm{E}\xi_x = \varphi(x)$, it is usually assumed that the majorant Neumann series that is obtained from the initial series by the changes $k(\cdot,\cdot) \to |k(\cdot,\cdot)|$, and $h(\cdot) \to |h(\cdot)|$ converges [19]. This is connected with the fact that, after the introduction of the indicator $\Delta_n = \Delta(x_n)$ of the nonbreak of a trajectory up to x_n, estimate (1.65) is expressed by the infinite series $\sum_{n=0}^{\infty} \Delta_n Q_n h(x_n)$, which is twice termwise averaged first over the break, i.e. over N for fixed $\{x_n\}$, and then over trajectory $\{x_n\}$. Such a procedure is used to define the quantity $\mathrm{E}\xi_x^2$. An analogous majorization is performed in the case of vector estimates for the systems of integral equations, where it may be especially heavy (see, e.g. [49]).

This work is largely devoted to the extension of the standard conditions of the unbiasedness of ξ_x [19] in the case when $p_n(x_n) \equiv p_n$, for which repeated averaging is performed quite simply first over the trajectory (for a fixed value of N) then over the break, i.e. over N.

Turning to the presentation of the new results, note that condition (1.66) with the boundedness of the operator K_1 ensures the validity of the equality

$$\mathrm{E}[\Delta_n Q_n h(x_n)] = K^n h(x), \qquad \forall n \geq 0. \tag{1.67}$$

Theorem 1.15. *Let*

$$\varphi(x) = \sum_{n=0}^{\infty} K^n h(x),$$

$$\sum_{n=0}^{\infty} |K^n h(x)| < +\infty, \tag{1.68}$$

$p_n(x_n) \equiv p_n, n = 0, 1, \ldots,$ *and*

$$\sum_{n=0}^{\infty} p_n = +\infty. \tag{1.69}$$

Then if

$$\mathrm{E}|\xi_x| < +\infty, \tag{1.70}$$

then

$$\mathrm{E}\xi_x = \varphi(x).$$

Proof. By (1.69), $\mathrm{P}(N < +\infty) = 1$ and, by (1.70), the Fubini theorem is applicable to $\mathrm{E}\xi_x$. Therefore,

$$\mathrm{E}\xi_x = \mathrm{E}\mathrm{E}(\xi_x|N) = \sum_{N=0}^{\infty} p_N \prod_{n=0}^{N-1}(1-p_n) \sum_{n=0}^{N} \Big[\prod_{k=0}^{n-1}(1-p_k)^{-1}\Big] K^n h(x). \tag{1.71}$$

If in (1.71) the equalities $K^n h(x)$ are replaced with their moduli and the order of summation is changed, then the series from (1.68) results. This fact makes it possible to change the order of summation in (1.71), which gives the desired result.

Note that condition (1.68) holds if the spectral radius $\rho(K) < 1$; however, as the proof of Theorem 1.15 shows, this condition is not sufficient for the unbiasedness of ξ_x even for constant probabilities p of the break. Hence, in those cases when one does not succeed in establishing the convergence of the above majorant Neumann series, it is expedient to numerically verify the stability of the sample mean of $|\xi_x|$; this is especially useful if a series of problems of a single type is being solved. If condition (1.68) is justified only on the heuristic level, then the stability of the mean

$$\mu_M = \frac{1}{M}\sum_{k=1}^{M}\sum_{n=0}^{\infty} \Delta_n^{(k)} Q_n^{(k)} h(x_n^{(k)})\Big|, \tag{1.72}$$

should be verified, where (k) is the number of independent realizations of the simulated Markov chain. It is clear that if $\sum_{n=0}^{\infty} |K^n h(x)| = +\infty$, then $\mathrm{P}(\mu_M \to +\infty) = 1$; thus, the stability of quantity (1.72) implies that condition (1.68) holds.

Note that if relation (1.67) is considered as the property of probabilistic approximation, then Theorem 1.15 can be considered to be a probabilistic analog of the known Lax theorem: convergence (in the sense of Kolmogorov's strong law of large numbers) follows from the probabilistic stability (1.70) and probabilistic approximation (1.67). Accordingly, the following simple statement is also of interest:

Theorem 1.16. *Let the solution to equation (1.64) be unique in L_∞, condition (1.66) hold, the Markov chain terminate with probability one, and* $\mathrm{E}|\xi_x| < C < +\infty, \forall x \in X$. *Then* $\mathrm{E}|\xi_x| = \varphi(x), \forall x \in X$.

Proof. It is obvious that

$$\xi_x = h(x) + q(x, x_1)\xi_{x_1}.$$

From here, on the basis of the known Tonelli theorem, one gets that the function $\mathrm{E}\xi_x$ satisfies equation (1.64); therefore, $\mathrm{E}\xi_x \equiv \varphi(x)$.

Under the hypotheses of Theorem 1.16, relation (1.66) can be considered as a condition that ensures the approximation of equation (1.64); hence, this theorem is a probabilistic analog of the Lax theorem with respect to the solution to equation (1.64) that is obtained by the Monte Carlo method. Note that the expansion of the solution into the Neumann series is not used here.

2. To solve the system of integral equations

$$\varphi_i(x) = \sum_{j=1}^{m} \int k_{ij}(x, x')\varphi_j(x')\mathrm{d}x' + h_i(x),$$

a vector estimate of the form of (1.65) with a matrix weight is constructed, and, after every transition $x_n \to x'$, the matrix weight of this estimate is multiplied by the matrix $\{k_{ij}(x_n, x')/p_n(x_n, x')\}$ [49]. Theorem 1.15 is extended to the vector case in the same form. Note that, in [67], the system of integral equations that are obtained by the mean-value theorem for the Lamé equations was solved by the Monte Carlo method. For a particular test problem, the inequality $\rho(K) < 1$ was established with some degree of certainty, and the stability of the statistical mean of $\mathrm{E}\xi_x$ was noted. On the basis of Theorem 1.15, these facts explain the satisfactoriness of the statistical estimates of the known solution that was obtained in [67] in the case when the majorant Neumann series seems to diverge.

Now consider the problem of estimating the quantity $\mathrm{E}\xi_x^2$, assuming for simplicity that $p_n(x, x') \equiv p(x, x')$ and $p_n \equiv p, n = 0, 1, \ldots$. The symbol K_p denotes the integral operator with the kernel $k^2(x, x')/p(x, x')$.

Theorem 1.17. *If conditions (1.66) - (1.69) are satisfied and*

$$\sum_{n=0}^{\infty} K_p^n\Big[h^2 + 2|h|\sum_{n=1}^{\infty}|K^n h|\Big](x) < +\infty, \tag{1.73}$$

then

$$\mathrm{E}\xi_x^2 = \sum_{n=0}^{\infty} K_p^n[h(2\varphi - h)]\,(x).$$

Proof. Since $\xi_x^2 \geq 0$, then the Fubini theorem is certainly applicable here. Averaging the expression

$$\xi_x^2 = \sum_{n=0}^{\infty} \Delta_n Q_n^2\Big[h^2(x_n) + 2h(x_n)\frac{k(x_n, x_{n+1})}{p(x_n, x_{n+1})}\Delta_{n+1}\xi_{x_{n+1}}\Big]$$

first over the trajectory and then over the break yields

$$\mathrm{E}\xi_x^2 = \sum_{n=0}^{\infty}(1-p)^{N-1}p\sum_{n=0}^{N}(1-p)^{-n}K_p^n\Big[h^2 + 2h\sum_{m=1}^{M}(1-p)^{-m}K^m h\Big](x). \tag{1.74}$$

Replacing the functions h and $K^n h$ with their moduli and performing the proper rearrangements of the order of summation result in the series of (1.73). Therefore, such rearrangements can be performed in (1.74), i.e. Theorem 1.17 is proved.

Theorem 1.17 is extended to the vector case as follows: The matrix integral operator defined by the formula from [49],

$$K_p\Psi(x) = \int_X \frac{k(x,x')\Psi(x')K^T(x,x')}{p(x,x')}\,\mathrm{d}x',$$

where $K(x,x')$ is the matrix of the kernels of the considered system of integral equations, is considered as K_p. Condition (1.73) takes on the form

$$\sum_{n=0}^{\infty} K_p^n\Big[H_1H_1^T + H_1\sum_{n=1}^{\infty}(K^nH)_1^T + \sum_{n=1}^{\infty}(K^nH)_1H_1^T\Big](x) < +\infty,$$

where, for example, H_1 denotes a column vector of the moduli of the elements of the vector H. If the latter inequality holds, the covariance matrix of the vector estimate ξ_x is defined as

$$\mathrm{E}(\xi_x\xi_x^T) = \sum_{n=0}^{\infty} K_p^n[H\Phi^T + \Phi H^T - HH^T]\,(x),$$

where Φ is a vector solution to the system. It is easy to see that a very significant weakening of the standard criterion that is connected with the replacement of all kernels of the system with their moduli occurs [49].

The linear functionals of the form $J_\Psi = (\psi, \varphi)$ are usually computed by the Monte Carlo method on the basis of the representation

$$J_\Psi = \mathrm{E}[q_0(x_0)\xi_{x_0}], \quad q_0(x_0) = \frac{\psi(x_0)}{\pi_0(x_0)},$$

where $\pi_0(x)$ is the probability density of the initial point x_0. Accordingly, the above statements are generalized simply. In particular, the condition $\sum_{n=0}^{\infty}|(\psi, K^n h)| < +\infty$ should be used instead of (1.68) and the relation

$$\sum_{n=0}^{\infty}\left(\psi^2/\pi_0, K_p^n\left[h^2+2|h|\sum_{n=1}^{\infty}|K^n h|\right]\right) < +\infty$$

should be used instead of condition (1.73).

3. Probabilistic moments of solutions to the problems with random parameters can be effectively computed by the Monte Carlo method on the basis of the corresponding randomization of the standard estimates [49], [19]. Actually, in this case, the Tonelli theorem is additionally used to go from the studied iterated integral to the double integral that expresses the expectation of the randomized estimate; hence, it is necessary to check the absolute stability (1.70) of such an estimate. It is clear that condition (1.68) takes on the form

$$\sum |\mathrm{M}K^n h(x)| < +\infty,$$

if $\mathrm{M}K_1^n h_1(x) < +\infty, \forall n \geq 0$; the symbol M denotes parametric averaging.

In conclusion, note that the method of antisymmetric sampling [28] is useful for attaining the absolute stability (1.70) of the considered estimates in the case of alternating signs. It can be especially effective for $p_n(x_n) \equiv p_n, n = 0, 1, ..., 0$, if the kernel $k(x, x')$ is substantially antisymmetric with respect to zero. The transition densities $p_n(x_n, x')$ must possess the corresponding symmetry, which is effectively realized by the replacement of some base random numbers $\{\alpha\}(0 < \alpha < 1)$ [19] with the numbers $\{1-\alpha\}$; in this case, along with ξ_x, the antisymmetric $\xi_x^{(1)}$ is constructed, which differs from ξ_x only by the indicated replacement. The quantity $\eta_x = (\xi_x + \xi_x^{(1)})/2$ may be absolutely stable in the case when ξ_x does not have this property. It is clear that statements of the types of 1.15 and 1.16 hold for η_x.

1.12. Approximate confidence intervals

To estimate the expected value of ξ, one usually uses a sequence $x_1, ..., x_N$ of independent samples of ξ obtained by the Monte Carlo technique. Then, by the law of large numbers,

$$\mathrm{E}\xi \approx \bar{x}_N = \frac{1}{N}\sum_{k=1}^{N} x_k \tag{1.75}$$

$$\mathrm{D}\xi = \sigma^2 \approx \tilde{\sigma}^2 = \frac{N}{N-1}\left(\frac{1}{N}\sum_{k=1}^{N} x_k^2 - \bar{x}_N^2\right).$$

To evaluate the error in the estimate of equation (1.75), a confidence interval is usually calculated which contains $\mathrm{E}\xi$ with probability $1-\varepsilon$, where $0 < \varepsilon \ll 1$. The central-limit theorem can be used to construct the confidence interval approximately, provided that $\mathrm{D}\xi < +\infty$. For N sufficiently large, the following asymptotic equality holds

$$P(|\bar{x}_N - I| < \beta\sigma/N^{1/2}) = P\left(\frac{\sqrt{N}}{\sigma}|\bar{x}_N - I| < \beta\right) \approx \frac{1}{\sqrt{2\pi}}\int_{-\beta}^{\beta} e^{-(x^2/2)}\mathrm{d}x = p(\beta).$$

Thus, setting $p(\beta) = 1-\varepsilon$, we obtain the desired confidence interval: $\bar{x}_N \pm \beta\sigma/(\Psi N)$. In practice, σ is usually unknown as well, and sample estimates $\tilde{\sigma}$ must be used for it.

In practice, the actual values of β are taken to be 1, 2 or 3, which correspond to the confidence levels $p(\beta) = 0.68, 0.95, 0.997$, respectively. When solving integral equations (see Section 1.3), the same trajectories are used to estimate a number of functionals for which the sample estimates are obtained as total contributions from various collisions. Clearly, it is not efficient in such calculations to estimate $\tilde{\sigma}$ by summation of x^2 over all trajectories. Therefore, the trajectories are divided up into a few groups each of which has M trajectories; from those data, the standard deviation of the random variable is estimated

$$\zeta_M = \frac{1}{M}\sum_{i=1}^{M} \xi_i,$$

where ξ_i are independent samplings of ξ. Let $\sigma_{M,S}$ be a statistical estimate of the quantity $\sigma_M = (\mathrm{D}\zeta_M)^{1/2}$ calculated for S groups ($N = MS$). Then the standard error of the quantity $\bar{x}_N$ can be estimated

$$\sigma(\bar{x}_N) = \frac{\sigma}{\sqrt{N}} \approx \frac{\tilde{\sigma}_{M,S}}{\sqrt{S}}.$$

This sample grouping allows us to substantiate the above confidence interval provided that M is large sufficiently for ζ_M to be practically the Gaussian random variable. Indeed, here we may use the Student confidence interval, which really coincides with the above one if $S \geq 30$.

Sometimes it is necessary to use, in the Monte Carlo method, estimates that have infinite variance. Let $F_\xi(y)$ be a distribution function of the random variable ξ and let

$$U(x) = \int_0^x y^2 \mathrm{d}F_\xi(y).$$

If the variance $\mathrm{D}\xi$ diverges as $\ln x$ (i.e., $U(x) \sim c \cdot \ln x$), then the random variable

$$\gamma_M = \left[\frac{M}{\ln M}\right]^{1/2} (\zeta_M - \mathrm{E}\xi),$$

is asymptotically normal $(0, d)$, where d is a standard deviation [22]. The standard error of the resultant estimate is given by

$$d^*(\zeta_{MS}) = d^* \left[\frac{\ln(MS)}{M \cdot S}\right]^{1/2},$$

which differs from the analogous quantity for the case of bounded variance by the factor $[\ln(MS)/\ln M]^{1/2}$. Although the variance of ζ_{MS} is infinite, the foregoing limit theorem enables us to construct approximately a confidence interval for $\mathrm{E}\xi$, as was explained for the normal distribution with $d = d(\zeta_{MS})$. Indeed, for large $N = M \cdot S$, the divergence of $\mathrm{D}\zeta_M$ is due to the tail of distribution which does not essentially affect the observation. This supposition agrees with the results of calculations. Therefore, the rate of probability convergence of estimate ζ_N is given by $t = [\ln N/(N)]^{1/2}$.

Chapter 2

Parametric estimates for solving problems of mathematical physics

2.1. Introductory information

1. Consider the three-dimensional Dirichlet problem for the Helmholtz equation

$$\Delta u + cu = -g, \quad u|_\Gamma = \psi, \tag{2.1}$$

in a domain $D \subset X_3$ with boundary Γ; moreover, $c < c^*$, where $-c^*$ is the first eigenvalue of the Laplace operator in the domain D, $r = (x, y, z) \in D$.

The regularity conditions on the functions g and ψ and the boundary Γ are assumed to be fulfilled; this ensures existence and uniqueness of a solution to problem (2.1) together with its probability representation and an integral representation by means of the spherical Green function [11], [20], [25].

The estimators of the Monte Carlo method considered further are connected with the so-called process of a 'random walk on spheres' in the domain D [59]. To describe the process, we introduce the following notation:
$\bar{D}$ is the closure of the domain D;
$d(P)$ is the distance from a point P to the boundary Γ;
Γ_ε is the ε-neighborhood of Γ, i.e. $\Gamma_\varepsilon = \{P \in \bar{D} : d(P) < \varepsilon\}$;
$S(P)$ is the maximal sphere centered at P and lying entirely in $\bar{D}, S(P) = \{Q \in \bar{D} : |Q - P| = d(P)\}$.

During a random walk on spheres, each consecutive point P_{k+1} is picked uniformly at random on the sphere $S(P_k)$ and the process terminates if the point enters Γ_ε.

Denote by $s_0(P, \varepsilon)$ that part of the sphere $S(P)$ which belongs to the set Γ_ε. Construct the sphere S_ε with radius ε and centre the tangency point of the boundary Γ and the sphere $S(P)$. Then the area of the part of $S(P)$ that lies entirely in S_ε equals $\pi\varepsilon^2$. Whence we infer the following estimate from below for the probability that the next point enters Γ_ε:

$$\frac{s_0(P,\varepsilon)}{4\pi d^2(P)} \geq \frac{\pi\varepsilon^2}{4\pi d^2(P)} \geq \frac{\varepsilon^2}{4d_{max}^2} = \nu(\varepsilon), \tag{2.2}$$

where d_{max} is the least upper bound of the radii of spheres lying entirely in the domain D. We assume that $|\psi|, |g| < C < +\infty$, and the domain D may be unbounded.

We now give a precise definition of the process of a random walk on spheres. Specify some Markov chain $\{P_n\}_{n=0,1,2,\dots,N}$ by the following characteristics:

$\pi(\mathbf{r}) = \delta(\mathbf{r} - P_0)$ is the density of the initial distribution (i.e. the chain starts at the point P_0);

$p(\mathbf{r}, \mathbf{r}') = \delta_\mathbf{r}(\mathbf{r}')$ is the density of transition from $\mathbf{r}$ to $\mathbf{r}'$, which represents the generalized three-dimensional density of the uniform probability distribution on the sphere $S(\mathbf{r})$;

$p_0(\mathbf{r})$ is the probability that the chain terminates; it is defined by the expression

$$p_0(\mathbf{r}) = \begin{cases} 0, & \mathbf{r} \notin \Gamma_\varepsilon; \\ 1, & \mathbf{r} \in \Gamma_\varepsilon; \end{cases}$$

N is the index of the terminal state.

As was already mentioned, the chain is called the *process of a random walk on spheres*. Obviously, the process can be written down as

$$P_n = P_{n-1} + \omega_n d(P_{n-1}), \quad n = 1, 2, \ldots,$$

where ω_n is the sequence of independent isotropic unit vectors.

The use of the process of a random walk on spheres for solving the Dirichlet problem was justified in [59]. On the grounds of the probability representation for a solution to the Dirichlet problem [25], the process was used in [16] for solving the Helmholtz equation with c = const ≤ 0 and in [15] with c = const $< c^*$. In [17], in the case c = const < 0 the corresponding algorithms were constructed and developed on the base of the integral representation related to the spherical Green function. Items 2-4 of this Section are devoted to developing the indicated approaches further and contain some new results on the possibilities of applying the estimators constructed earlier.

2. The probability $p_1(\mathbf{r})$ that the chain terminates after a consecutive transition from $\mathbf{r}$ to $\mathbf{r}'$ can be estimated as follows:

$$p_1(\mathbf{r}) = \int_{\Gamma_\varepsilon} p(\mathbf{r}, \mathbf{r}') p_0(\mathbf{r}') \, d\mathbf{r}' \geq \nu(\varepsilon). \tag{2.3}$$

Hence, the average number $q(P_0, \varepsilon)$ of transitions in the chain of a random walk on spheres does not exceed the quantity $\nu^{-1}(\varepsilon)$. Moreover, for a wide class of boundaries Γ, on the grounds of the renewal theory, the logarithmic estimate [20], [46], [58]

$$q(P_0, \varepsilon) \leq C|\ln \varepsilon| \tag{2.4}$$

has been constructed, holding at least for convex domains. We point out that, for some types of nonconvex domains with angular points, estimators of the form (2.4) were obtained in [46].

Estimate (2.4) becomes heuristically evident if we consider the density of the distribution of the centers of spheres. It is well known [59] that the process of a random walk on spheres converges with probability 1 to the boundary; therefore, it practically suffices to consider the density near a planar boundary. Denote by x the distance to the planar boundary. The similarity argument shows that, to within a constant multiplier, the density $f(x)$ of the distribution of the centers of spheres with respect to x must be close to x^{-1}. Whence we readily infer estimate (2.4) [17].

However, for a planar boundary in the three-dimensional case, the renewal theory gives the following more precise result:

Theorem 2.1. *If Γ is a plane, then*

$$q(P_0, \varepsilon) = \frac{\ln(\varepsilon/d_0) - 1}{\ln 2 - 1},$$

where $d_0 = d(P_0)$

Proof. Represent the quantity $\ln d_N$ as follows [46]:

$$\ln d_N = \ln d_0 + \sum_{k=1}^{N} \ln \frac{d_k}{d_{k-1}}.$$

For a planar boundary, the quantities $\ln(d_k/d_{k-1})$ are obviously independent and identically distributed; in particular, in the three-dimensional case the representations $d_k = 2d_{k-1}\alpha_k$ hold, where $\{\alpha_k\}$ are independent random variables identically distributed in the interval (0, 1), and

$$\mathrm{E}\Big(\ln \frac{d_{k+1}}{d_k}\Big) = \ln 2 - 1.$$

In accordance with the renewal theory (see, for instance, [6]), we have

$$q(P_0, \varepsilon) = \frac{\ln(\varepsilon/d_0) + \mathrm{E}\gamma(\ln \varepsilon)}{\ln 2 - 1},$$

where $\gamma(\ln \varepsilon)$ is the random variable of the jump of the quantity $\ln\, d_N$ over the level $\ln \varepsilon$; i.e. $\gamma(\ln \varepsilon) = \ln d_N - \ln \varepsilon < 0$. It is obvious that in this case the quantity $-\gamma(\ln \varepsilon)$ has standard exponential distribution; i.e. $\mathrm{E}\gamma(\ln \varepsilon) \equiv -1$. □

Theorem 2.1 gives an upper bound for the value $q(P_0, \varepsilon)$ for a convex domain [59], [46].
3. *Use of the probability representation.* In the case $c = \text{const} < c^*$ the probability representation for a solution to problem (2.1) has the form [25]

$$u(\mathbf{r}_0) = \mathrm{E} \int_0^\tau e^{ct} g(\xi(t))dt + \mathrm{E}[e^{c\tau}\psi(\tau))],$$

where $\xi(t)$ is the diffusion process corresponding to the Laplace operator and starting at the point $\mathbf{r}_0$ and τ is the moment of the first exit of the process from the domain D. By the strong Markov property, we have

$$\begin{aligned} u(\mathbf{r}_0) &= \mathrm{E}\sum_{i=0}^{\infty} \int_{\tau_i}^{\tau_{i+1}} e^{ct} g(\xi(t))dt + \mathrm{E}\Big[\psi(\xi(\tau))\prod_{i=0}^{\infty} e^{c(\tau_{i+1}-\tau_i)}\Big] \\ &= \sum_{i=0}^{\infty} \mathrm{E}\Big[e^{c\tau_i} \int_0^{\tau_{i+1}-\tau_i} e^{ct} g(t+\tau_i)dt\Big] + \mathrm{E}\Big[\psi(\xi(\tau))\prod_{i=0}^{\infty} e^{c(\tau_{i+1}-\tau_i)}\Big], \end{aligned}$$

where τ_i is the moment of the first occurrence of the process $\xi(t)$ corresponding to the random walk on spheres $\{\mathbf{r}_n\}, n = 0, ..., N$, on the surface of the *i*th sphere; moreover, $\mathbf{r}_n \in \Gamma_\varepsilon$.

Denote by P some point of the boundary Γ which is nearest to $\mathbf{r}_N$. On using the iterated averaging together with the integral representation of a solution to problem (2.1) at the center of the ball, it is easy to obtain [16], [15]

$$u(\mathbf{r}_0) \approx u_\varepsilon(\mathbf{r}_0) = \mathrm{E}\eta_\varepsilon^{(0)},$$

where

$$\eta_\varepsilon^{(0)} = \sum_{i=1}^{N} \Big[\prod_{j=0}^{i-1} s(c,d_j)\Big] \int_{D(\mathbf{r}_i)} G(\rho;c,d_i)g(\rho)\mathrm{d}\rho + \Big[\prod_{j=0}^{N-1} s(c,d_j)\Big]\psi(P). \tag{2.5}$$

Here $d_j = d(\mathbf{r}_j), D(\mathbf{r}_i)$ is the ball with radius d_i and centre at the point $\mathbf{r}_i$, $G(\rho;c,d)$ is the spherical Green function, and

$$s(c,d) = \begin{cases} d\sqrt{c}/\sin(d\sqrt{c}), & c \geq 0, \\ d\sqrt{|c|}/\sin(d\sqrt{|c|}), & c \leq 0. \end{cases}$$

The possibility of applying the iterated averaging becomes obvious if we take account of the fact that the quantities $\tau_{i+1} - \tau_i$ are conditionally independent (for fixed $\{\mathbf{r}_i\}$) and the distribution of the quantity $\tau_{i+1} - \tau_i$ is independent of the position of the point $\xi(\tau_{i+1})$ on the sphere $S(\mathbf{r}_i)$.

Further, each of the integrals involved in expression (2.5) can be estimated [17] by means of its random node ρ_i which it is reasonable to choose in $D(\mathbf{r}_i)$ with density proportional to $G(\rho;c,d_i)$. For $c \leq 0$, this density will be considered in Item 4. Moreover, instead of $\eta_\varepsilon^{(0)}$, we obtain the estimator

$$\eta_\varepsilon = \sum_{i=o}^{N} \Big[\prod_{j=0}^{i-1} s(c,d_j)\Big]\, g(\rho_i)F_{\mathbf{r}_i} + \Big[\prod_{j=0}^{N-1} s(c,d_j)\Big]\psi(P), \tag{2.6}$$

where

$$F_{\mathbf{r}} = \int_{D(\mathbf{r})} G(\rho;c,d(\mathbf{r}))d\rho \leq C_1 d^2(\mathbf{r}).$$

The iterated averaging shows that

$$\mathrm{E}\eta_\varepsilon = \mathrm{E}\eta_\varepsilon^{(0)} = u_\varepsilon(\mathbf{r}_0).$$

We now discuss the question of uniform (in ε) boundedness of the value of the variance $\mathrm{D}\eta_\varepsilon$ determining the mean square error of the resultant estimator by the Monte Carlo method which is obtained by averaging the realizations of η_ε. We settle the question in the case $c > 0$, since in the case $c \leq 0$ it is easily resolved by the method of integral representations (see Item 4).

Theorem 2.2. *Under the conditions of the problem, for* $0 < c < c^*/2$ *and* $|\psi|, |g| < C_1 < +\infty$ *we have*

$$\mathrm{D}\eta_\varepsilon < C < +\infty.$$

Proof. It is obvious that $|\eta_\varepsilon| \leq |\eta_\varepsilon^{(1)}|$, where $\eta_\varepsilon^{(1)}$ corresponds to $\psi \equiv g \equiv C_1$. Furthermore, the quantity $\eta^{(1)}$ represents the conditional expectation of the corresponding Wiener estimator $\xi^{(1)}$; i.e. $\mathrm{D}\eta^{(1)} \leq \mathrm{D}\xi^{(1)}$; moreover, $\mathrm{E}(\eta_\varepsilon^{(1)})^2 \leq \mathrm{E}(\eta^{(1)})^2$. It is known [25] that $\mathrm{D}\xi^{(1)} < +\infty$ for $g \equiv 0$ and $c < c^*/2$. Thus, Theorem 2.2 holds in the case $g \equiv 0$. Now, denote the quantity $\eta^{(1)}$ corresponding to the given $g(\cdot)$ by $\eta^{(1)}(g)$ and look at the relation

$$\prod_{j=0}^{n} s(c,d_j) = \prod_{j=0}^{n-1} s(c,d_j) + [-1+s(c,d_n)] \prod_{j=0}^{n-1} s(c,d_j) = \dots$$
$$\dots = 1 + \sum_{i=0}^{n} \Big[\prod_{j=0}^{i-1} s(c,d_j) \Big] [-1 + s(c,d_j)].$$

Since $d_{\max}\sqrt{c} < \pi$, it follows that $-1+s(c,d_i) \geq C_2 d_i^2$. In consequence, $\eta^{(1)}(g) \leq C_3\eta^{(1)}(0)$, and Theorem 2.2 is completely proven. □

Theorem 2.3. *If $c < c^*$ and the first-order derivatives of a solution to problem (2.1) are bounded in D then*

$$|u(\mathbf{r}) - \mathrm{E}\eta_\varepsilon| \leq \text{const} \cdot \varepsilon, \quad \mathbf{r} \in D, \quad \varepsilon > 0.$$

Proof. The following relation is obvious:

$$|\zeta_\varepsilon - \eta_\varepsilon| \leq \text{const } Q_N^{(0)} \varepsilon,$$

where $Q_N^{(0)}$ is the weight corresponding to the estimator of a random walk on spheres for the given value of $c, \varepsilon = 0$, and $g \equiv 0$; and ζ_ε is the unbiased estimator for $u(\mathbf{r})$ related to the use of the value $u(\mathbf{r}_N)$ instead of $\psi(P)$. Since $|\mathrm{E}(\zeta_\varepsilon - \eta_\varepsilon)| \leq \mathrm{E}|\zeta_\varepsilon - \eta_\varepsilon|$, Theorem 2.3 is thus proven. □

Theorem 2.3 demonstrates that under the condition $d_{\max} < +\infty$, a solution with uniformly bounded derivatives is unique.

On the other hand, the quantity $\mathrm{D}\eta_\varepsilon$ must, as a rule, be unbounded for $c > c_0/2$, where c_0 is a solution to the equation

$$s^2(c_0/2, d_{\max}) = s(c^*, d_{\max}),$$

since in this case $c_0 > c^*$, and $s^2(c,d) \geq s(c^*,d)$. Moreover, here $\mathrm{D}\eta_\varepsilon = +\infty$ for ε sufficiently small, because of the above-analyzed correspondence between the estimators in the random walk on spheres and the diffusion process and because of the relations

$$c^*(D - \Gamma_\varepsilon) \to c^*(D) = c^*, \quad \varepsilon \to 0.$$

Calculation shows [50] that the relation $\mathrm{D}\eta_\varepsilon < +\infty$ in the case $c^*/2 < c < c_0/2$ is quite possible, but a rigorous proof is still unavailable.

4. *Use of integral representations.* Consider the following three-dimensional Dirichlet problem:

$$\Delta u - cu = -g, \quad u|_\Gamma = \psi, \tag{2.7}$$

in a domain D with boundary $\Gamma; c = \text{const} \geq 0$. The regularity conditions on g, ψ, and Γ are assumed to be fulfilled which ensure existence and uniqueness of a solution to problem (2.7) together with the possibility of representing a solution by means of Green's function for a ball. Write down the integral equation for the function $u(\mathbf{r})$ [17]:

$$u(\mathbf{r}) = \int_D k(\mathbf{r},\mathbf{r}')u(\mathbf{r}')\,\mathrm{d}\mathbf{r}' + h(\mathbf{r}), \tag{2.8}$$

where

$$k(\mathbf{r},\mathbf{r}') = \begin{cases} \dfrac{d\sqrt{c}}{\sinh(d\sqrt{c})}\delta_{\mathbf{r}}(\mathbf{r}'), & \mathbf{r} \notin \Gamma_\varepsilon; \\ 0, & \mathbf{r} \in \Gamma_\varepsilon; \end{cases}$$

$$h(\mathbf{r}) = \begin{cases} \dfrac{1}{4\pi}\displaystyle\int \dfrac{\sinh[(d-|\mathbf{r}'-\mathbf{r}|)\sqrt{c}]}{|\mathbf{r}'-\mathbf{r}|\sinh(d\sqrt{c})} g(\mathbf{r}')\,d\mathbf{r}', & \mathbf{r} \notin \Gamma_\varepsilon; \\ u(\mathbf{r}), & \mathbf{r} \in \Gamma_\varepsilon. \end{cases}$$

Here $d = d(\mathbf{r})$, $\delta_{\mathbf{r}}(\mathbf{r}')$ is the generalized density corresponding to the uniform distribution on the sphere $S(\mathbf{r})$. Since $d\sqrt{c}/\sinh(d\sqrt{c}) \leq 1$ for $c \geq 0$, condition (2.2) implies

$$\int_D\int_D k(\mathbf{r},\mathbf{r}')k(\mathbf{r}',\mathbf{r}'')\,d\mathbf{r}''d\mathbf{r}' \leq \int_{D-\Gamma_\varepsilon} \delta_{\mathbf{r}}(\mathbf{r}')\Big(\int_D \delta_{\mathbf{r}'}(\mathbf{r}'')\,d\mathbf{r}''\Big)\,d\mathbf{r}' = \int_{D-\Gamma_\varepsilon} \delta_{\mathbf{r}}(\mathbf{r}')\,d\mathbf{r}' \leq 1-\nu(\varepsilon).$$

Consequently, as regards the space L_∞ which is natural here, we have

$$||K^2|| \leq 1-\nu(\varepsilon) < 1.$$

This fact guarantees convergence for the Neumann series for equation (2.8), and thereby the applicability of the Monte Carlo method. Moreover, it follows that the solution to the initial differential problem is a unique solution to equation (2.8). Here we have no problem with traces, since the operator K leaves invariant the set of functions only with discontinuities of the first kind on the inner boundary of the set Γ_ε.

For estimation, we can use the relation

$$u(\mathbf{r}_0) = \mathrm{E}\zeta, \quad \zeta = h(\mathbf{r}_0) + \sum_{n=1}^{N} Q_n h(\mathbf{r}_n),$$

where $\{\mathbf{r}_n\}$ is the chain of an isotropic random walk on spheres terminating in Γ_ε (in the case $\varepsilon > 0$ it is reasonable to call this chain the ε-spherical process), and the weights are determined by the formulas

$$Q_0 = 1, \quad Q_n = Q_{n-1}\frac{d_{n-1}\sqrt{c}}{\sinh(d_{n-1}\sqrt{c})} \leq Q_{n-1}, \quad d_n = d(P_n), \quad n = 1,2,\ldots.$$

We now recall that we have temporarily introduced the unreal supposition: the solution $u(\mathbf{r})$ is known in Γ_ε. However, in place of exact values of $u(\mathbf{r})$ in Γ_ε, we can use approximate values, for instance, by taking them from the nearest points of the boundary; i.e. we can set

$$u(\mathbf{r}) \approx \psi(\mathbf{r}^*), \quad \mathbf{r} \in \Gamma_\varepsilon, \quad |\mathbf{r}-\mathbf{r}^*| = d(\mathbf{r}).$$

As a result, we obtain some biased estimator ζ_ε whose mean value $u_\varepsilon(\mathbf{r}_0)$ differs from $u(\mathbf{r}_0)$ by a quantity of order ε. Indeed, if we consider the expression for the difference of the estimators ζ and ζ_ε, then we obtain

$$|u-u_\varepsilon| \leq |\mathrm{E}\{Q_N[u(\mathbf{r}_N) - \psi(\mathbf{r}_N^*)]\}| \leq A\varepsilon,$$

where A is some constant which is finite because of boundedness of the derivatives of the solution in the domain D (use is made of the relation $Q_N \leq 1$).

It is rather simple to establish uniform boundedness for the quantity $\mathrm{D}\zeta_\varepsilon$.

Theorem 2.4. *Under the conditions of the problem we have* $\mathrm{D}\xi_\varepsilon < C + \infty, \quad \forall\varepsilon > 0$.

Proof. The quantity ζ_ε increases monotonically as $c \to 0$, and for $c = 0$ we have $Q_N \equiv 1$ and (see, for instance, [50])

$$\mathrm{E}\zeta_\varepsilon^2 = h_\varepsilon(2u_\varepsilon - h_\varepsilon) + K_\varepsilon(\mathrm{E}\zeta_\varepsilon^2).$$

Consequently,

$$\mathrm{E}\zeta_\varepsilon^2 \le K_\varepsilon(\mathrm{E}\zeta_\varepsilon^2) + C|h_\varepsilon|,$$

and thereby the quantity $\mathrm{E}\zeta_\varepsilon^2$ is dominated by a solution to the equation $\chi = K_\varepsilon\chi + Ch_\varepsilon^{(1)}$ which corresponds to the initial problem after the substitutions $g \to C_0|g|$ and $\psi \to C_0|\psi|$.

Thus, the quantity $\mathrm{D}\zeta_\varepsilon$ is uniformly bounded if the solution remains bounded after the substitutions, provided that $c \ge 0$ and we have the above-indicated regularity of the conditions of the problem. The theorem is proven. □

As in Item 3, we evaluate the integral that represents $h(\mathbf{r})$ for $\mathbf{r} \notin \Gamma_\varepsilon$ by using the Monte Carlo method for a single random node. As a result, instead of ζ_ε we obtain a random variable $\zeta_{\varepsilon,1}$. The iterated averaging shows that $\mathrm{E}\zeta_{\varepsilon,1} = u_\varepsilon(\mathbf{r}_0)$. Applying an analogous averaging for the variance, it is easy to show that $\mathrm{D}\zeta_{\varepsilon,1}$ is bounded owing to the boundedness of $g(\mathbf{r})$.

Evaluate the number of arithmetic operations necessary for achieving the required error ε in estimation of a solution. As was shown earlier, the number of spheres for the trajectories to terminate in Γ_ε is of the order of the quantity $\ln\varepsilon$. On the other hand, for the probability error of estimation to be of the order of ε, one must simulate $C\varepsilon^{-2}$ trajectories. Hence,

$$R_\varepsilon \sim C_n|\ln\varepsilon|/\varepsilon^2, \tag{2.9}$$

where C_n is the number of arithmetic operations per one sphere in the space of n dimensions. At large n the quantity C_n depends on n linearly [46]. We note that relation (2.9) was first indicated in [17].

5. From what was said above, it is clear that the construction of estimators which uses a random walk on spheres by the conditional averaging of Wiener estimators is most effective within the frame of the probability representation. On the other hand, the use of an integral equation (in the case when the spectral radius of the corresponding operator is less than unity) gives an opportunity to construct various weighted modifications connected for example with an anisotropic random walk.

Observe that in the case when the averaging of the Wiener estimator is ineffective or even impossible, the martingale technique presented in [58] and developed in detail in [20] may be very useful.

2.2. Solving the Helmholtz equation with a complex parameter

1. Consider the three-dimensional Dirichlet problem for the Helmholtz equation

$$\Delta u + cu = -g, \quad u|_\Gamma = \psi \tag{2.10}$$

in a domain $D \subset X_3$ with boundary Γ. Here c is a complex parameter, $c = a + bi$. We assume that the functions g and ψ and the boundary Γ satisfy some regularity conditions that guarantee existence and uniqueness of a solution to problem (2.10) as well as the integral representation for a solution by means of the Green function for a ball inscribed in D (see Section 2.1).

The estimates under consideration below are constructed on the base of the process of a random walk on spheres in the domain D (see Section 2.1). We note that the domain D may be unbounded but the following relations are assumed valid: $d \leq d_{\max} < \infty$ and $|\psi|, |g| < C < \infty$.

In the process $\{\mathbf{r}_n\}, n = 0, 1, \ldots, N$, of a random walk on spheres, the next point $\mathbf{r}_{n+1}$ is chosen uniformly at random on the surface of the sphere $S(\mathbf{r}_n)$. The process terminates if the point enters Γ_ε, i.e. $\mathbf{r}_N \in \Gamma_\varepsilon$. The estimate $\mathrm{E}N \leq C(\Gamma)|\ln \varepsilon|$ holds for a sufficiently broad class of Γ. In Item 1 of the present Section, we extended the estimators of the Monte Carlo method which were constructed earlier for a real parameter to the case of a complex parameter [54] by using the method of [50].

As in the real case (see Section 2.1), a solution to problem (2.10) satisfies the integral relation

$$u_1(\mathbf{r}) = \int\limits_{\bar{D}} k(\mathbf{r}, \mathbf{r}') u_1(\mathbf{r}')\, d\mathbf{r}' + h(\mathbf{r}), \tag{2.11}$$

where

$$k(\mathbf{r}, \mathbf{r}') = \begin{cases} q(c, d)\delta_{\mathbf{r}}(\mathbf{r}'), & \mathbf{r} \notin \Gamma_\varepsilon, \\ 0, & \mathbf{r} \in \Gamma_\varepsilon, \end{cases} \qquad q(c, d) = \frac{d\sqrt{c}}{\sin(d\sqrt{c})}$$

$$h(\mathbf{r}) = \begin{cases} \int\limits_{D(\mathbf{r})} G(\rho; c, d) g(\rho) d\rho, & \mathbf{r} \notin \Gamma_\varepsilon \\ u(\mathbf{r}), & \mathbf{r} \in \Gamma_\varepsilon \end{cases}$$

Here $\delta_{\mathbf{r}}(\mathbf{r}')$ is a generalized density corresponding to the uniform probability distribution on the sphere $S(\mathbf{r})$ and $G(\rho; c, d)$ is the Green function on the ball $D(\mathbf{r})$ bounded by the sphere $S(\mathbf{r})$. The following representation is valid:

$$\begin{aligned} \int\limits_{D(\mathbf{r})} G(\rho; c, d) g(\rho)\, d\rho &= \frac{1}{4\pi} \int\limits_{\Omega} d\omega \int\limits_0^d x\Big(1 - \frac{x}{d}\Big) \frac{q(c, d)}{q(c, d - x)} g(x, \omega)\, dx \\ &= \frac{d^2}{6} \mathrm{E}\Big\{\frac{q(c, d)}{q(c, d - \nu)} g(\nu, \omega)\Big\}, \end{aligned} \tag{2.12}$$

where ω is the isotropic direction and ν is the random variable distributed in the interval $(0, d)$ with density $6x(1 - x/d)\, d^{-2}$. The results below are connected with the properties of the function $q(c, d)$ of the complex variable c. They are illustrated in the next assertion:

Lemma 2.1. *The following inequalities are valid for* $c = a + bi$:

$$|q(c, d)| \leq |q(a, d)|; \tag{2.13}$$

$$|q(c,d)| \le |q(bi,d)|, \quad a<0; \tag{2.14}$$
$$|q(c,d)| \le |q(c,d-x)|, \quad a<0, \quad 0 \le x \le d. \tag{2.15}$$

Proof. The following well-known representation is valid for the function $(\sin z)/z$:

$$\frac{\sin z}{z} = \prod_{n=1}^{\infty}\left(1-\frac{z^2}{n^2\pi^2}\right). \tag{2.16}$$

Consequently,

$$\left|\frac{\sin d\sqrt{c}}{d\sqrt{c}}\right| = \left|\prod_{n=1}^{\infty}\left(1-\frac{d^2c}{n^2\pi^2}\right)\right| \ge \prod_{n=1}^{\infty}\left|1-\frac{d^2a}{n^2\pi^2}\right| = \left|\frac{\sin d\sqrt{a}}{d\sqrt{a}}\right|.$$

Whence we obtain (2.13). Moreover, the relation

$$\left|1-\frac{d^2c}{n^2\pi^2}\right| = \left(1+\frac{d^2|a|}{n^2\pi^2}\right)^2 + \left(\frac{bd^2}{n^2\pi^2}\right)^2 \ge \left|1-\frac{id^2b}{n^2\pi^2}\right|^2 \tag{2.17}$$

holds for $a<0$, whence (2.14) ensues. Inequality (2.15) is easy to obtain from the first half of (2.17). □

Below, we also need the representation

$$\left|\frac{z}{\sin z}\right|^2 = \frac{|z|^2}{\sin^2(\mathrm{Re}(z)) + \mathrm{sh}^2(\mathrm{Im}(z))}. \tag{2.18}$$

Now, we shall construct and substantiate the necessary probability representations. Henceforth, the symbol c^* stands for the absolute value of the first eigenvalue of the Laplace operator on the domain D.

Lemma 2.2. *If $Re(c) = a < c^*$, then the Neumann series $\sum_{n=0}^{\infty} K^n h$ for equation (2.11) remains convergent after replacing the kernel $k(\cdot,\cdot)$ and the function $h(\cdot)$ with their moduli.*

Proof. By (2.13), Lemma 2.2 follows from the earlier proven corresponding assertion for a real parameter (see, for instance, [50]). □

Theorem 2.5. *If $Re(c) < c^*$, then there is a unique bounded solution to equation (2.11) given by the corresponding Neumann series and coinciding with a solution to problem (2.10).*

Proof. If $u_0(\mathbf{r})$ is the Neumann series and K_0 is the integral operator of equation (2.11) for $0 < c = c_0 < c^*$ and $\psi \equiv g \equiv C_0$, then $u_0(\mathbf{r}) \ge C_0$ and $\lim_{n\to\infty} K_0^n u_0(\mathbf{r}) = 0$ [50]. Therefore, if $u(\mathbf{r})$ is a solution to equation (2.11) and $|u(\mathbf{r})| < C_1$, then from (2.13) we obtain

$$u = \sum_{i=0}^{n} K^i h + K^{n+1} u;$$

moreover, $|K^{n+1}u| < C_1 C_0^{-1} K_0^{n+1} u_0$. □

Using the above assertion, we can represent a solution to problem (2.10) as the expectation of the standard 'estimator by collisions' connected with a random walk on spheres. Randomizing the value $h(\mathbf{r})$ according to (2.12), we can rewrite the estimator as

$$\xi = \sum_{n=0}^{N} \Big[\prod_{i=0}^{n-1} q(c, d_i) \Big] g(\nu_n, \omega_n) \frac{d_n^2 q(c, d_n)}{6q(c, d_n - \nu_n)} + \Big[\prod_{n=0}^{N-1} q(c, d_n) \Big] u(\mathbf{r}_N)$$

It is clear that

$$|q(c, d_n)/q(c, d_n - \nu_n)| < C < +\infty.$$

The general theory of the estimator by collisions (see Section 1.3), with Lemma 2.2 taken into account, claims that $\mathrm{E}\xi = u(\mathbf{r})$. Moreover, using the estimator of the quantity $\mathrm{E}\xi^2$ for a real parameter (see Section 2.1) and (2.13), we infer that $\mathrm{E}|\xi^2| < +\infty$ if $\mathrm{Re}(c) < c^*/2$.

Similarly, we establish that the replacement of $u(\mathbf{r}_N)$ with $\psi(P(\mathbf{r}_N))$ at the nearest point of the boundary (denote the so-modified estimator by ξ_ε) leads to the ε-displacement; i.e. $|u(\mathbf{r}) - \mathrm{E}\xi_\varepsilon| < C\varepsilon$, provided that the modulus of the gradient of a solution is uniformly bounded in Γ_ε. Here the quantity $\mathrm{E}|\xi_\varepsilon^2|$ is bounded together with $\mathrm{E}|\xi^2|$.

2. The above results enable us to implement the Fourier transform while solving the mixed problem for a parabolic equation by the Monte Carlo method.

Consider the problem

$$u_t - \Delta u = f(\mathbf{r}, t), \quad u|_{t=0} = \varphi(\mathbf{r}), \quad u|_\Gamma = v(\mathbf{r}, t), \tag{2.19}$$

with $v(\mathbf{r}, 0)|_\Gamma = \varphi(\mathbf{r})$. We suppose that $f(\mathbf{r}, t) \equiv v(\mathbf{r}, t) \equiv u(\mathbf{r}, t) \equiv 0$ for $t < 0$.

First, we assume that $f \equiv \varphi \equiv 0$. In this case the Fourier transform of u is a solution to the boundary value problem

$$\Delta U - isU = 0, \quad U|_\Gamma = V, \tag{2.20}$$

where $V = V(\mathbf{r}, s) = \int_{-\infty}^{+\infty} u(t) e^{-ist} dt$ is the Fourier transform of v. Alongside the regularity conditions outlined at the begining of Item 1, we suppose that the integral of $|v(\mathbf{r}, t)|$ over the time $v_0(\mathbf{r})$ converges; moreover, $v_0(\mathbf{r}) < C < +\infty, \mathbf{r} \in \bar{D}$. We suppose that the function u possesses the same properties. It is well known that, under these suppositions, the corresponding Fourier transforms possess similar properties. Also, we suppose that the derivatives u_x, u_y and u_z are bounded in magnitude by functions depending only on t and having convergent integrals. This in particular guarantees uniqueness for a solution to problem (2.20) even over an unbounded domain, provided that $d_{\max} < +\infty$ (see Section 2.1). Consequently, we can calculate, an approximate solution to the problem by estimating

$$J(\mathbf{r}, t) = \frac{1}{2\pi} \int_{-\infty}^{+\infty} U(\mathbf{r}, s) e^{ist} ds = u(\mathbf{r}, t)$$

while solving problem (2.20) by the Monte Carlo method simultaneously for different values of s as indicated in Item 1. Due to the conditions imposed on the space derivatives, the ε-displacement of the individual estimators transforms into the ε-displacement of the estimator for the integral. In constructing the corresponding quadrature formula, the following assertion turns out to be useful (the symbol $\zeta_\varepsilon(\mathbf{r}, s)$ denoting the probability estimator for a solution to problem (2.20) which was considered in Item 1):

Lemma 2.3. *If $v(\mathbf{r}, t) \in C_0^k(-\infty < t < +\infty)$ and $\int_{-\infty}^{+\infty} |v_t^{(k)}(\mathbf{r}, t)| dt < C_0 < +\infty, \mathbf{r} \in \Gamma$, then the following inequality is valid:*

$$|U_\varepsilon^2(\mathbf{r},s)| = |\mathrm{E}\zeta_\varepsilon(\mathbf{r},s)|^2 \le \mathrm{E}|\zeta_\varepsilon^2(\mathbf{r},s)| \le \frac{Cd^2(\mathbf{r})|s|e^{-d(\mathbf{r})\sqrt{2|s|}}}{(1+|s|)^{2k}} \tag{2.21}$$

Proof. By (2.13), all weight factors in the estimator $\zeta_\varepsilon(\mathbf{r},s)$ do not exceed unity in magnitude, and by (2.18) we obtain

$$\mathrm{E}|\zeta_\varepsilon^2(\mathbf{r},s)| \le \frac{C_1|q^2(-is,d)|}{(1+|s|)^{2k}} = \frac{C_1}{(1+|s|)^{2k}}\frac{d^2|s|}{\sin^2(d\sqrt{|s|/2}) + sh^2(d\sqrt{|s|/2})},$$

since $\sqrt{-is} = (1-i)\sqrt{s/2}$. Here the function $|V|$ in the conditions of the lemma is estimated by the factor $(1+|s|)^{-2k}$. □

Observe that the quantity $J(\mathbf{r},t)$ can be estimated by the double randomization method [19], [50] with the help of the equality

$$J_\varepsilon(\mathbf{r},t) = \frac{1}{2\pi}\mathrm{E}\chi_\varepsilon, \quad \chi_\varepsilon = \frac{\zeta_\varepsilon(\mathbf{r},\eta)}{p_\eta(\eta)}e^{i\eta t},$$

where $p_\eta(s)$ is a suitable probability density in $[0, \infty)$. In particular, we can use a density proportional to the function $[\mathrm{E}|\zeta_\varepsilon^2(\mathbf{r},s)|]^{1/2}$ which satisfies the inequality

$$\mathrm{E}|\chi_\varepsilon|^2 \le \Big(\int_{-\infty}^{+\infty} [\mathrm{E}|\eta_\varepsilon^2(\mathbf{r},s)|]^{1/2}\mathrm{d}s\Big)^2.$$

If we use the double randomization then the algorithm has cost of order $m|\ln\varepsilon|\varepsilon^{-2}$, where m is the number of points at which the function $u(\mathbf{r},t)$ is estimated.

On the other hand, using Lemma 2.3, we can construct the following effective determinate quadrature formula for the integral $J(\mathbf{r},t)$:

$$J \approx \tilde{J} = \sum_{k=1}^{M} p_k U(\mathbf{r},s_k), \quad p_k \ge 0, \quad \sum_{k=1}^{M} p_k = 1,$$

with error of order $M^{-\alpha}, \alpha > 0$. Since $\tilde{J}^2 \le \sum_{k=1}^{M} p_k U(\mathbf{r},s_k)$, the mean squared probability error of the statistic estimator for the quantity J has the form $\sigma/\sqrt{S}$, where S is the volume of a sample in the Monte Carlo method. Equating the determinate and probability errors, i.e., putting $S = O(\varepsilon^{-2})$ and $M = O(\varepsilon^{-1/\alpha})$, we obtain an algorithm whose cost is estimated asymptotically by a quantity of order $|\ln\varepsilon|\varepsilon^{-(2+1/\alpha)}$ for every m.

The usefulness of the method increases essentially, since we can estimate the functionals of the form $\int_{-\infty}^{+\infty} u(\mathbf{r},t)\psi_i(t)dt$ simultaneously almost without increasing the cost of the algorithm (basing on the well-known Parseval formula). Executing additional randomization of the initial point $\mathbf{r}$, we can also estimate the functionals of the form $\int_{-\infty}^{+\infty} u(\mathbf{r},t)\psi_i(\mathbf{r},t)\mathrm{d}\mathbf{r}\mathrm{d}t$ immediately. Differentiating $J(\mathbf{r},t)$ with respect to t under the integral sign, we can construct simultaneously implemented estimators for the corresponding derivatives of a solution and the indicated functionals of these derivatives (under some well-known additional regularity conditions).

It is obvious that the estimator for the Fourier transform of a solution to the general problem (2.19) has the form $\zeta_\varepsilon(\mathbf{r},s)+\eta_\varepsilon(\mathbf{r},s)$, where η_ε corresponds to problem (2.19) with

$v(\mathbf{r},t) \equiv 0$, i.e. it is determined by the first summand in the expression for the estimator ξ of Item 1 with $c = -is$ and $g(\mathbf{r},s) = F(\mathbf{r},s) - \varphi(\mathbf{r})$. As for the function $f(\mathbf{r},t)$, we make the same suppositions as those above for the function $v(\mathbf{r},t)$. Here it is reasonable to use separately some numerical method for estimating the part of the solution to problem (2.19) which is a solution to the same problem with the zero boundary conditions in the initial ball $S(\mathbf{r})$ (which can be taken to be not maximal). It is obvious that the estimator $\eta_\varepsilon^{(1)}(\mathbf{r},s)$ corresponding to the remaining part of the solution is obtained from $\eta_\varepsilon(\mathbf{r},s)$ by nullifying the initial summand which is connected with $S(\mathbf{r})$. By analogy with Lemma 2.3 while taking (2.15) into account, we obtain the estimate

$$|\eta_\varepsilon^{(1)}(\mathbf{r},s)| \le C_1 d^2(\mathbf{r})|s| \exp(-d(\mathbf{r})\sqrt{2|s|})\eta^{(0)}(\mathbf{r}), \tag{2.22}$$

where $\eta^{(0)}(\mathbf{r})$ is the estimator for a solution to the homogeneous Dirichlet problem in the domain D for the Poisson equation with a constant right-hand side sufficiently large in magnitude. It is well known (see Section 2.1) that

$$\mathrm{E}(\eta^{(0)}(\mathbf{r}))^2 < C < +\infty, \quad \mathbf{r} \in D.$$

On the other hand, the inequality $|\eta_\varepsilon(\mathbf{r},s)| \le C_2(1+|s|)^{-k}$ holds for $\varphi \equiv 0$ and $f \in C_0^k$, which also enables us to apply the above method of calculation for $k \ge 2$. If we choose the point (ν_0, ω_0) in the initial sphere according to a density proportional to the function $|G(\nu,\omega;-is,d)|$ then, as is easy to show, we can additionally obtain the factor $(1+|s|)^{-1}$ in the estimate for the quantity $|\eta_\varepsilon(\mathbf{r},s)|$; i.e. this method applies to the case $k = 1$ as well.

For estimating the solution of the above mentioned problem in a ball let us take into consideration a parameter d^0, such that the ball with the centre in the point x^0 and the radius d^0 belongs to the domain D :

$$\omega_t = \Delta\omega, \quad \omega|_{t=0} = \varphi(\mathbf{r}), \quad \omega|_{S_0} = 0, \quad x \in \{|x - x^0| \le d^0\},\ t \ge 0. \tag{2.23}$$

We obtain the solution of problem (2.23) by solving the Cauchy problem for the heat equation with the same initial conditions:

$$\omega_t^{(1)} = \Delta\omega^{(1)}, \quad \omega^{(1)}|_{t=0} = \varphi(\mathbf{r}), \quad t \ge 0, \tag{2.24}$$

and then we estimate the solution of the problem:

$$\omega_t^{(2)} = \Delta\omega^{(2)}, \quad \omega^{(2)}|_{t=0} = 0, \quad \omega^{(2)}|_{S_0} = -\omega^{(1)}|_{S_0}, \quad x \in \{|x - x^0| \le d^0\},\ t \ge 0 \tag{2.25}$$

in the centre of the ball. The Fourier transform of boundary problem (2.25) has the following form:

$$\Delta W^{(2)} - isW^{(2)} = 0, \quad W^{(2)}|_{S_0} = -W^{(1)}|_{S_0}, \quad x \in \{|x-x^0| \le d^0\},\ s \in (-\infty,+\infty) \tag{2.26}$$

where $W^{(1)}$ and $W^{(2)}$ are the respective Fourier transforms of $\omega^{(1)}$ and $\omega^{(2)}$.

The solution of problem (2.24) has the following representation [11]:

$$\omega^{(1)}(x,t) = \frac{1}{(2\sqrt{\pi t})^3} \int_{-\infty}^{+\infty} \varphi(\xi) e^{-\sum_{i=0}^{3}(\xi_i - x_i)^2/4t} \, d\xi. \tag{2.27}$$

It is obvious that this integral can be represented as the expectation:

$$\omega^{(1)}(x,t) = \mathrm{E}\varphi(\xi), \quad \xi = (\xi_1, \xi_2, \xi_3),$$

where ξ_i are independent normally distributed random variables with the parameters $[x_i, \sqrt{2t}]$.

The solution of problem (2.23) is represented as a sum of solutions of (2.24) and (2.25) in the point x^0. The same method is used for calculation of the estimator to the solution of problem (2.25) as well as for problem (2.20). The corresponding value of the boundary condition is estimated by calculation of the Fourier transform of integral (2.27) in a point on the sphere $S_0 = \{|x - x^0| = d^0\}$, which is chosen as the end point of the random walk on spheres for problem (2.26). In this case, the random walk on spheres estimator is determined by one point to be chosen uniformly on sphere S_0. Here we need the additional condition: $\mathrm{E}\varphi^2(\xi) < +\infty$.

Now, consider a problem like (2.19) for the more general equation $u_t - \Delta u + \lambda u = f$. The above results extend automatically to the case $\lambda = \text{const} > 0$, since the substitution of $c = -is - \lambda$ for $c = -is$ decreases the quantity $|q(c,d)|$ according to (2.14). Under some additional conditions, we can extend these results to the case of a varying parameter: $0 \le \lambda(\mathbf{r}) \le \lambda_0$. To this end, in line with [20], we rewrite the equation $\Delta U - (is + \lambda)U = -g$ as the equation

$$\Delta U - (is + \lambda_0)U = -g - (\lambda_0 - \lambda)U$$

to which there corresponds the following integral equation with one more integral over the ball $D(\mathbf{r})$ [50], [20]:

$$U_1(\mathbf{r}, s) = \frac{q(-\lambda_0 - is, d)}{4\pi d^2} \int\limits_{S(\mathbf{r})} U_1(\mathbf{r}(\chi), s)\mathrm{d}\chi + \int\limits_{D(\mathbf{r})} G(\rho; -\lambda_0 - is, d)[U_1(\rho, s) + g(\rho, s)]\mathrm{d}\rho. \tag{2.28}$$

Randomizing the equation as proposed in [50], we obtain the relation

$$\begin{aligned} U_1(\mathbf{r}, s) &= \Big(1 - \frac{\lambda_0 d^2}{6}\Big)\frac{q(-\lambda_0, d)}{1 - \lambda_0 d^2/6}\frac{q(-\lambda_0 - is, d)}{q(-\lambda_0, d)} \\ &\times \frac{1}{4\pi d^2}\int\limits_{S(\mathbf{r})} U_1(\mathbf{r}(\chi), s)\mathrm{d}\chi + \frac{\lambda_0 d^2/6}{4\pi}\int\limits_{\Omega}\mathrm{d}\omega \int\limits_0^d \Big(\frac{d^2}{6}\Big)^{-1} x\Big(1 - \frac{x}{d}\Big) \\ &\times \frac{q(-\lambda_0 - is, d)}{q(-\lambda_0 - is, d - x)}\Big[\frac{\lambda_0 - \lambda(\rho)}{\lambda_0}U_1(x, \omega; s) + g(x, \omega; s)\Big]\mathrm{d}x. \end{aligned}$$

Here both ratios of the values of the function q are less than unity by Lemma 2.1; hence, in this case, for $\lambda_0 < c^*$, we can use the weighted algorithm of a random walk on spheres and balls [50] for which the probability of entering a ball equals $\lambda_0 d^2/6$. The variance of the corresponding estimator is bounded if the condition $\lambda_0 < 0.488c^*$ holds. Such an algorithm is rather simple to implement as well as to extend to the multidimensional case (see Item 3 below). However, the weighted modification of the algorithm considered in [20] turns out sometimes to be more effective; this modification is based on the following randomized representation for (2.28):

$$U_1(\mathbf{r}, s) = q(-\lambda_0, d)\frac{q(-\lambda_0 - is, d)}{q(-\lambda_0, d)}\frac{1}{4\pi d^2}\int\limits_{S(\mathbf{r})} U_1(\mathbf{r}(\chi), s)\mathrm{d}\chi$$

$$+ [1 - q(-\lambda_0, d)] \int\limits_{D(\mathbf{r})} p(\mathbf{r}, \rho) \frac{G(\rho; -\lambda_0 - is, d)}{G(\rho; -\lambda_0, d)} \Big[\frac{g(\rho)}{\lambda_0} + \frac{\lambda_0 - \lambda(\rho)}{\lambda_0} U_1(\rho, s)\Big] d\rho,$$

where the probability density $p(\mathbf{r}, \rho)$ has the form [20]

$$p(\mathbf{r}, \rho) = \frac{\lambda_0 \mathrm{sh}[d - |\mathbf{r} - \rho|\sqrt{\lambda_0})]}{4\pi|\mathbf{r} - \rho|[\mathrm{sh}(d\sqrt{\lambda_0}) - d\sqrt{\lambda_0}]}, \quad \rho \in D(\mathbf{r}).$$

We can demonstrate that $|G(\rho; -\lambda_0 - is, d)/G(\rho; -\lambda_0, d)| \leq 1$. This relation is an obvious consequence of the well-known probability representation for the Green function for the Helmholtz equation in the case of a complex parameter c [31]. Observe that according to the last randomization, the transition to a ball occurs with probability

$$1 - q(\lambda_0, d) = 1 - d\sqrt{\lambda_0}/\mathrm{sh}(d\sqrt{\lambda_0}).$$

It is easy to see that the algorithms for the variable $\lambda(\mathbf{r})$ lack the factors for which effective exponential estimators like (2.21) can be constructed. Consequently, these algorithms are in general inapplicable to the case of $\varphi \equiv 0$.

3. In the general n-dimensional case, in relation (2.11) we have [11]

$$q(c, d) = \frac{(d\sqrt{c}/2)^{(n-2)/2}}{\Gamma(n/2) J_{(n-2)/2}(d\sqrt{c})},$$

where $J_{(n-2)/2}(\cdot)$ is the Bessel function. The quantity $h(\mathbf{r})$ can be randomized as follows:

$$h(\mathbf{r}) = \frac{d^2}{2n} \int p_0(\mathbf{r}, \rho)[G(\rho; c, d)/G(\rho; 0, d)] g(\rho) d\rho, \tag{2.29}$$

where (for $n > 2$)

$$p_0(\mathbf{r}, \rho) = 2nd^{-2} G(\rho; 0, d) = \frac{2n}{(n-2) d^2 \omega_n} \Big(\frac{1}{(\rho - \mathbf{r})^{n-2}} - \frac{1}{d^{n-2}}\Big), \quad |\rho - \mathbf{r}| \leq d.$$

Here $\omega_n = 2(\sqrt{\pi})^n/\Gamma(n/2)$ is the surface of the n-dimensional unit sphere. It is easy to verify that $\int p_0(\mathbf{r}, \rho) d\rho = 1$. For $n = 2$ we have

$$p_0(\mathbf{r}, \rho) = 4d^{-2} G(\rho; 0, d) = \frac{2}{\pi} \ln \frac{d}{|\rho - \mathbf{r}|}, \quad |\rho - \mathbf{r}| \leq d.$$

The ratio of the values of the function G in (2.29) is bounded, since the Green functions with different values of c have poles of the same order at $\rho = 0$. According to (2.29), the function $d^2 q(c, d)/[6q(c, d - \nu)]$ in the expression for ξ of Item 1 is replaced with the function $d^2 G(\rho; c, d)/[2nG(\rho; 0, d)]$, where ρ is a random point in $D(\mathbf{r})$ distributed with probability density $p_0(\mathbf{r}, \rho)$.

Lemma 2.1 is valid in the n-dimensional case, since

$$\psi(z) = \Gamma(n/2) J_{(n-2)/2}(z)/(z/2)^{(n-2)/2}$$

is representable in the form (2.16) with the positive roots of the function $J_{(n-2)/2}(\cdot)$ substituted for $\{n\pi\}$ (it is well known that $\psi(0) = 1$ and that all roots of the Bessel function are real and their set is symmetric about the origin [4]). Consequently, all results of Item 1 extend to the n-dimensional case.

Since the asymptotic equality

$$|J_{(n-2)/2}(z)| = O(|z|^{-1}e^{|z|})$$

holds at large $|z|$ and $|\arg z| < \pi/2 - \delta$ [4], relation (2.21) is valid in the general case with the factor $|s|$ of the exponential function replaced with $|s|^{(n-1)/2}$.

Further, the first randomization of (2.28) is constructed with (2.29) taken into account. The variance of the corresponding estimator is finite if $\lambda_0 < \gamma_n c^*$ [50]; in particular, $\gamma_4 = 0.465$ and $\gamma_{10} = 0.336$. The second randomization preserves its form upon passage to the n-dimensional case; however, we should put

$$p(\mathbf{r}, \rho) = \frac{G(\rho; -\lambda_0, d)}{1 - q(-\lambda_0, d)}, \quad \rho \in D(\mathbf{r}).$$

Using the probability representation for the Green function [31], we can easy demonstrate that $\int G(\rho; -\lambda_0, d)d\rho = 1 - q(\lambda_0, d)$; i.e. $p(\mathbf{r}, \rho)$ represents a probability density.

2.3. Solution of boundary value problems of the second and third kinds

1. Consider the following boundary value problem in a boundary domain D $\subset R^n$ with boundary Γ :

$$\Delta u - cu = -g, \ \mathbf{r} \in D, \qquad \alpha u + \beta\frac{\partial u}{\partial \mathbf{l}} = \varphi, \ \mathbf{r} \in \Gamma, \tag{2.30}$$

where $\alpha(\mathbf{r}) \geq 0$, $\beta(\mathbf{r}) \geq 0$, and $c \geq 0$. The vector field $\mathbf{l}$ on Γ satisfies the following conditions: $(\mathbf{l}(\mathbf{r}), \mathbf{n}(\mathbf{r})) > \theta_0 > 0$, $||\mathbf{l}|| = 1$, $\mathbf{r} \in \Gamma$, where $\mathbf{n}(\mathbf{r})$ is the outward normal vector to Γ at $\mathbf{r}$. We suppose that α and β are bounded and do not vanish simultaneously: $\alpha + \beta \geq 1$.

We suppose that $\alpha, \beta, g, \varphi$, and the boundary Γ satisfy the regularity conditions that guarantee existence and uniqueness of a solution to the problem [11].

2. In this Item, we construct an integral equation whose unique solution coincides with the exact solution to the above problem. Afterwards, using the standard technique of the Monte Carlo method, we shall construct an estimator for the solution [53].

To begin with, we obtain a difference approximation to the boundary condition of (2.30) in an ε-neighborhood $\Gamma_\varepsilon = \{\mathbf{r} \in D : d(\mathbf{r}) < \varepsilon\}$ of the boundary; here $d(\mathbf{r})$ is the distance to Γ. Given $\mathbf{r} \in \Gamma_\varepsilon$, denote by $\mathbf{b} = \mathbf{b}(\mathbf{r}) \in \Gamma$ a point such that the ray with vertex $\mathbf{b}$ and direction vector $-\mathbf{l}(\mathbf{b})$ passes through $\mathbf{r}$. We assume that, for ε sufficiently small, such a point can be chosen uniquely among the points of the boundary which lie in some neighborhood of $\mathbf{r}$. Using the Taylor series expansion of a solution along the direction of $\mathbf{l}(\mathbf{r})$, we can obtain the following expression:

$$u(\mathbf{r}) = \frac{\beta + \alpha\Delta}{\beta + \alpha\Delta + \alpha h}u(\mathbf{r} - h\mathbf{l}) + \frac{h}{\beta + \alpha\Delta + \alpha h}\varphi(\mathbf{b}) + \phi(\mathbf{r}), \tag{2.31}$$

where $\alpha = \alpha(\mathbf{b})$, $\beta = \beta(\mathbf{b})$, and $\Delta = |\mathbf{b} - \mathbf{r}|$.

To derive (2.31), let us use the following Taylor expansion:

$$
\begin{aligned}
u(\mathbf{b}) &= u(\mathbf{r}) + \frac{\partial u(\mathbf{r})}{\partial \mathbf{l}}|\mathbf{r}-\mathbf{b}| + \frac{\partial^2 u}{\partial \mathbf{l}^2}(\mathbf{b} - \theta_1|\mathbf{r}-\mathbf{b}|\mathbf{l})|\mathbf{r}-\mathbf{b}|^2/2, \\
u(\mathbf{r}-h\mathbf{l}) &= u(\mathbf{r}) - \frac{\partial u(\mathbf{r})}{\partial \mathbf{l}}h + \frac{\partial^2 u}{\partial \mathbf{l}^2}(\mathbf{r} - \theta_2 h\mathbf{l})\frac{h^2}{2}, \\
\frac{\partial u(\mathbf{b})}{\partial \mathbf{l}} &= \frac{\partial u(\mathbf{r})}{\partial \mathbf{l}} + \frac{\partial^2 u}{\partial \mathbf{l}^2}(\mathbf{r} - \theta_3|\mathbf{r}-\mathbf{b}|\mathbf{l})|\mathbf{r}-\mathbf{b}|.
\end{aligned}
$$

Here $\theta_i = \theta_i(\mathbf{r}) \in (0,1)$, $i = 1,2,3$. Denote the residual members by $U_i(\mathbf{r})$, $i = 1,2,3$. Further,

$$
\alpha(\mathbf{b})(u(\mathbf{r}) + \frac{\partial u(\mathbf{r})}{\partial \mathbf{l}}\Delta + U_1(\mathbf{r})) + \beta(\mathbf{b})(\frac{\partial u(\mathbf{r})}{\partial \mathbf{l}} + U_3) = \varphi(\mathbf{b}),
$$

$$
\alpha(\mathbf{b})u(\mathbf{r}) + \frac{\partial u(\mathbf{r})}{\partial \mathbf{l}}(\alpha(\mathbf{b})\Delta + \beta(\mathbf{b})) + \alpha(\mathbf{b})U_1(\mathbf{r}) + \beta(\mathbf{b})U_3 = \varphi(\mathbf{b}),
$$

$$
\frac{\partial u(\mathbf{r})}{\partial \mathbf{l}} = (u(\mathbf{r}) - u(\mathbf{r}-h\mathbf{l}))/h + U_2/h.
$$

Therefore,

$$
\alpha u(\mathbf{r}) + ((u(\mathbf{r}) - u(\mathbf{r}-h\mathbf{l}))/h + U_2/h)(\alpha\Delta + \beta) + \alpha U_1(\mathbf{r}) + \beta U_3 = \varphi(\mathbf{b}),
$$

$$
(\alpha(h+\Delta) + \beta)u(\mathbf{r}) - (\alpha\Delta + \beta)u(\mathbf{r}-h\mathbf{l}) + (\alpha\Delta+\beta)U_2 + \alpha h U_1(\mathbf{r}) + \beta h U_3 = \varphi(\mathbf{b})h.
$$

As a result, we obtain (2.31) with

$$
\phi(\mathbf{r}) = -\frac{1}{\beta + \alpha h + \alpha\Delta}\left(\alpha h U_1 + (\alpha\Delta + \beta)U_2 + \beta h U_3\right).
$$

Assuming that the second-order partial derivatives of a solution to (2.30) are bounded in Γ_ε, we can determine the error order

$$
\max_{\mathbf{r}\in\Gamma_\varepsilon} |\phi(\mathbf{r})| \le C_1 \varepsilon h + C_2 h^2. \tag{2.32}
$$

By analogy with the Dirichlet problem [17], [50], using (2.31) we write the integral equation that holds for a solution to the original differential problem:

$$
u_1(\mathbf{r}) = \int_D k(\mathbf{r},\mathbf{r}')u_1(\mathbf{r}')d\mathbf{r}' + v(\mathbf{r}), \tag{2.33}
$$

where

$$
k(\mathbf{r},\mathbf{r}') = \begin{cases} s(c,d)\delta_{S(\mathbf{r})}(\mathbf{r}'),\ \mathbf{r} \notin \Gamma_\varepsilon, & s(c,d) = \dfrac{(d\sqrt{c}/2)^{\frac{n-2}{2}}}{\Gamma(\frac{n}{2})I_{\frac{n-2}{2}}(d\sqrt{c})}, \\ q(\mathbf{r})\delta_{\mathbf{r}-h\mathbf{l}}(\mathbf{r}'),\ \mathbf{r} \in \Gamma_\varepsilon, & q(\mathbf{r}) = \dfrac{\beta + \alpha\Delta}{\beta + \alpha\Delta + \alpha h}, \end{cases}
$$

$$
v(\mathbf{r}) = \begin{cases} \int_{D(\mathbf{r})} G(\rho;c,d)g(\rho)d\rho,\ \mathbf{r} \notin \Gamma_\varepsilon, \\ \dfrac{h}{\beta + \alpha\Delta + \alpha h}\varphi(\mathbf{r}) + \phi(\mathbf{r}),\ \mathbf{r} \in \Gamma_\varepsilon. \end{cases} \tag{2.34}
$$

Here $\delta_{S(\mathbf{r})}$ is the generalized density corresponding to the uniform probability distribution on the sphere $S(\mathbf{r}) = \{\mathbf{r}' \in \bar{\mathrm{D}} : |\mathbf{r} - \mathbf{r}'| = d(\mathbf{r})\}$, $d = d(\mathbf{r})$; $\delta_{\mathbf{r}-h\mathbf{l}}$ is the generalized density corresponding to the transition from $\mathbf{r}$ to $\mathbf{r} - h\mathbf{l}(\mathbf{b})$; $\alpha = \alpha(\mathbf{b})$, $\beta = \beta(\mathbf{b})$, $\Delta = |\mathbf{b} - \mathbf{r}|$, and $\mathbf{b} = \mathbf{b}(\mathbf{r})$, where $\mathbf{r} \in \Gamma_\varepsilon$; and $G(\rho; c, d)$ is the Green function for the ball $\mathrm{D}(\mathbf{r})$ bounded by the sphere $S(\mathbf{r})$. The value $v(\mathbf{r})$ in $\mathrm{D} \setminus \Gamma_\varepsilon$ can be randomized as follows:

$$v(\mathbf{r}) = \frac{d^2}{2n} \int\limits_{\mathrm{D}(\mathbf{r})} p_0(\mathbf{r}, \rho)(G(\rho; c, d)/G(\rho; 0, d))g(\rho)d\rho, \tag{2.35}$$

where (for $n > 2$) the probability density $p_0(\mathbf{r}, \rho)$ equals

$$2nd^{-2}G(\rho; 0, d) = \frac{2n}{(n-2)d^2\omega_n}\Big(\frac{1}{(\rho - \mathbf{r})^{n-2}} - \frac{1}{d^{n-2}}\Big), \quad |\rho - \mathbf{r}| \le d. \tag{2.36}$$

Here $\omega_n = 2(\sqrt{\pi})^n/\Gamma(n/2)$ is the area of a unit n-dimensional sphere. For $n = 2$ we have

$$p_0(\mathbf{r}, \rho) = 4d^{-2}G(\rho; 0, d) = \frac{2}{\pi} \ln \frac{d}{|\rho - \mathbf{r}|}, \quad |\rho - \mathbf{r}| \le d.$$

The ratio of the values of G in (2.35) is bounded, since the Green functions for different values of c have poles of the same order at $\rho = 0$.

Suppose temporarily that the error function ϕ is known. In this case, substituting the exact solution $u(\mathbf{r})$ of (2.30) for $u_1(\mathbf{r})$ in (2.33), we come obviously to an identity. We now demonstrate convergence of the Neumann series for (2.33). Introduce the notation

$$\gamma = \max_{\mathbf{b} \in \Gamma} \frac{\beta(\mathbf{b})}{\alpha(\mathbf{b})h + \beta(\mathbf{b})} \le 1, \quad q_0 = q(c, \varepsilon), \quad \nu(\mathbf{r}, \varepsilon) = \frac{\mu(\mathbf{r}, \varepsilon)}{\omega_n d^{n-1}(\mathbf{r})} < 1, \quad \mathbf{r} \in D \setminus \Gamma_\varepsilon$$

where $\mu(\mathbf{r}, \varepsilon)$ is the area of the part of $S(\mathbf{r})$ that lies in Γ_ε.

In what follows, for simplicity we suppose that if $\mathbf{r} \in \Gamma_\varepsilon$ then $\mathbf{r} - h\mathbf{l} \in \mathrm{D} \setminus \Gamma_\varepsilon$. To this end, we may assume that $h \ge C\varepsilon$, with C a sufficiently large constant.

Lemma 2.4. *If one of the conditions $\alpha(\mathbf{r}) \ge \alpha_0 > 0$ and $c > 0$ is satisfied, then $||K^2||_{L_\infty} < 1$.*

Proof. It is well known [50] that $q(c, d) \le q_0 < 1$ for $c > 0$ and $d \ge \varepsilon$. Moreover, if $\alpha(\mathbf{r}) \ge \alpha_0 > 0$, then $\gamma < 1$. Given $\mathbf{r} \in D \setminus \Gamma_\varepsilon$, we have

$$\int\limits_{\mathrm{D}}\int\limits_{\mathrm{D}} k(\mathbf{r}, \mathbf{r}')k(\mathbf{r}', \mathbf{r}'')d\mathbf{r}'d\mathbf{r}'' \le q_0 \int\limits_{\mathrm{D}} \int\limits_{\mathrm{D}\setminus\Gamma_\varepsilon} \delta_{S(\mathbf{r})}(\mathbf{r}')k(\mathbf{r}', \mathbf{r}'')d\mathbf{r}'d\mathbf{r}''$$

$$+ q_0 \int\limits_{\mathrm{D}}\int\limits_{\Gamma_\varepsilon} \delta_{S(\mathbf{r})}(\mathbf{r}')k(\mathbf{r}', \mathbf{r}'')d\mathbf{r}'d\mathbf{r}'' \le q_0^2(1 - \nu(\mathbf{r}, \varepsilon)) + q_0\gamma\nu(\mathbf{r}, \varepsilon) < 1.$$

Furthermore, the relation

$$\int\limits_{\mathrm{D}}\int\limits_{\mathrm{D}} k(\mathbf{r}, \mathbf{r}')k(\mathbf{r}', \mathbf{r}'')d\mathbf{r}'d\mathbf{r}'' \le \gamma q_0 < 1$$

holds for $h \ge C\varepsilon$ and $\mathbf{r} \in \Gamma_\varepsilon$. Finally,

$$\int_D \int_D k(\mathbf{r},\mathbf{r}')k(\mathbf{r}',\mathbf{r}'')\mathrm{d}\mathbf{r}'\mathrm{d}\mathbf{r}'' < \delta < 1, \quad \forall \mathbf{r} \in D.$$

□

Lemma 2.4 guarantees convergence of the Neumann series for (2.33). Moreover, a solution to the original differential problem (2.30) is a unique solution to (2.33); weak solutions are implied as well.

We can slightly weaken the supposition that $\mathbf{r} \in \Gamma_\varepsilon$ implies $\mathbf{r} - h\mathbf{l} \in \mathrm{D} \setminus \Gamma_\varepsilon$, requiring for instance that $\mathbf{r}$ leaves Γ_ε in at most m steps (see the next Item). Then we can prove that $||K^m||_{L_\infty} < 1$ under the hypotheses of the lemma.

3. Consider the following modification of the process of a random walk on spheres (see Section 2.1). In a random walk inside $\mathrm{D} \setminus \Gamma_\varepsilon$, the next point $\mathbf{r}_{k+1}$ is chosen uniformly at random on the sphere $S(\mathbf{r}_k)$. Entering Γ_ε, we determine $\mathbf{b}_k \in \Gamma$ and calculate the next point by the formula $\mathbf{r}_{k+1} = \mathbf{r}_k - h\mathbf{l}(\mathbf{b}_k)$, where the jump length h is not less than $C\varepsilon$. Henceforth we suppose that, after reflection from the boundary into the interior of D at a distance h along a nontangential vector to the boundary, the corresponding segment lies entirely in the domain for h sufficiently small. Suppose that $R(\mathbf{r})$ is the probability that the trajectory terminates at a point $\mathbf{r} \in \mathrm{D}$. Define two ways of terminating the trajectory: (1) the trajectory terminates in a walk inside $\mathrm{D} \setminus \Gamma_\varepsilon$, and (2) entering Γ_ε, the trajectory terminates with some probability and reflects with the complementary probability.

We discuss the question of conditions under which the mean number $\mathrm{E}L$ of transitions in the chain is finite and determine the asymptotic order of this value and the mean number $\mathrm{E}\eta$ of reflections. For a broad class of boundaries Γ, the following logarithmic estimate is well known (see Section 2.1) for the chain of a random walk on spheres with termination in Γ_ε :

$$\mathrm{E}L \leq C_0 |\ln(\varepsilon/d_0)|, \tag{2.37}$$

where d_0 is the distance from the initial point of the chain to the boundary. M.E. Muller [59] showed that the mean number of transitions in the chain of a random walk on spheres depends linearly on n for n large: $\mathrm{E}L \sim n$. Using these results, we obtain a corresponding estimate in our case.

Consider the first model when the trajectory terminates in $\mathrm{D} \setminus \Gamma_\varepsilon$ with probability $R(\mathbf{r}) = 1 - s(c,d)$. In what follows, we use the following auxiliary problem:

$$\Delta p_1 - cp_1 = 0, \quad p_1|_\Gamma = 1. \tag{2.38}$$

Lemma 2.5. *Suppose that $c > 0$ and the trajectory of the chain of a random walk on spheres with reflection in the boundary terminates with the following probability: $R(\mathbf{r}) = 1 - s(c,d)$, $\mathbf{r} \in \mathrm{D} \setminus \Gamma_\varepsilon$ and $R(\mathbf{r}) = 0$, $\mathbf{r} \in \Gamma_\varepsilon$. If $\varepsilon = h^2$ and*

$$P_1 = \min_{\mathbf{b} \in \Gamma} \left| \frac{\partial p_1(\mathbf{b})}{\partial \mathbf{l}(\mathbf{b})} \right| > 0,$$

then $\mathrm{E}\eta \asymp 1/h$, and $\mathrm{E}L \asymp |\ln h|/h$ as $h \to 0$.

Proof. It is clear that if the process starts at $\mathbf{r}_0 \notin \Gamma_\varepsilon$ then the probability equals $p_1(\mathbf{r}_0) + O(\varepsilon)$ [50] that the trajectory does not terminate until entering Γ_ε. Therefore, under the condition $P_1 > 0$ and h sufficiently small, the probability R that the trajectory terminates between the time of reflection in the boundary and the time of the next hitting of Γ_ε satisfies the inequality

$$R \geq P_1 h + O(h^2) + O(\varepsilon).$$

Hence, $\mathrm{E}\eta = O(1/h)$ for h sufficiently small. The assertion on the order of $\mathrm{E}L$ is obvious from the estimate for $\mathrm{E}\eta$ and (2.37). □

Now, consider the second model in which the trajectory, entering Γ_ε, terminates with probability $R(\mathbf{r})$ and reflects in the boundary with probability $1 - R(\mathbf{r})$.

Lemma 2.6. *Suppose that* $\min\limits_{\mathbf{b}\in\Gamma}(\alpha(\mathbf{b})) = \alpha_0 > 0$ *and the process terminates with the following probability:* $R(\mathbf{r}) = 0$, $\mathbf{r} \in \mathrm{D} \setminus \Gamma_\varepsilon$, *and* $R(\mathbf{r}) = 1 - q(\mathbf{r})$, $\mathbf{r} \in \Gamma_\varepsilon$. *If* $\varepsilon = h < 1$, *then* $\mathrm{E}\eta \asymp 1/h$ *and* $\mathrm{E}L \asymp n/h$ *as* $h \to 0$ *and* $n \to \infty$.

Proof. Since the coefficients $\alpha(\mathbf{r})$ and $\beta(\mathbf{r})$ are bounded, we have

$$1 - q(\mathbf{r}) \geq \frac{\alpha_0 h}{\max\alpha + \max\beta},$$

whence the assertion of the lemma follows. □

Consider the problem for the equation $\Delta u - cu = -g$ with mixed boundary conditions of the form

$$\begin{cases} \alpha u + \beta \dfrac{\partial u}{\partial \mathbf{l}} = \varphi_1, & \mathbf{r} \in \Gamma_1, \\ u = \varphi_2, & \mathbf{r} \in \Gamma_2, \end{cases} \tag{2.39}$$

where Γ_1 is an open subset of Γ, and $\Gamma_2 = \Gamma \setminus \Gamma_1$ is nonempty. To solve such a problem, we construct chains that terminate in $\Gamma_{2,\varepsilon}$ with probability $R(\mathbf{r}) = 1$ [35]. By analogy with [35], we introduce the following auxiliary problem:

$$\Delta p_2(\mathbf{r}) = 0, \quad p_2(\mathbf{r}) = 0,\ \mathbf{r} \in \Gamma_1, \quad p_2(\mathbf{r}) = 1,\ \mathbf{r} \in \Gamma_2.$$

Lemma 2.7. *Considering the problem for the equation* $\Delta u - cu = -g$ *with the boundary conditions (2.39), assume that the process terminates with the following probability:* $R(\mathbf{r}) = 0$, $\mathbf{r} \in D \setminus \Gamma_{2,\varepsilon}$ *and* $R(\mathbf{r}) = 1$, $\mathbf{r} \in \Gamma_{2,\varepsilon}$. *If* $\varepsilon = h^2$ *and*

$$P_2 = \min_{\mathbf{b}\in\Gamma_1} \left| \frac{\partial p_2(\mathbf{b})}{\partial \mathbf{l}(\mathbf{b})} \right| > 0,$$

then $\mathrm{E}\eta \asymp 1/h$ *and* $\mathrm{E}L \asymp |\ln h|/h$ *as* $h \to 0$.

Proof. is given in [35]. □

4. For all $r = \mathbf{r}_0 \in \mathrm{D}$, write down the estimator for the solution in recurrent form

$$\xi_i = Q(\mathbf{r}_i)\xi_{i+1} + v_i, \quad Q(\mathbf{r}_i) = \begin{cases} \dfrac{s(c,d_i)}{1-R(\mathbf{r}_i)}, & \mathbf{r}_i \in D\setminus\Gamma_\varepsilon, \\ \dfrac{q(\mathbf{r}_i)}{1-R(\mathbf{r}_i)}, & \mathbf{r}_i \in \Gamma_\varepsilon. \end{cases} \tag{2.40}$$

With probability $1 - R(\mathbf{r}_i)$ the estimator ξ_{i+1} is calculated at the next point $\mathbf{r}_{i+1}$ of the chain and the estimator of the free term is added to it. The chain terminates with the complementary probability $R(\mathbf{r}_i)$. Here $v_i = v(\mathbf{r}_i)$, given by (2.34), is estimated in $\mathrm{D}\setminus\Gamma_\varepsilon$ at random according to (2.36) and is calculated in Γ_ε exactly. Obviously, the inequality $R(\mathbf{r}) < 1$ must hold if the weight q is nonzero.

Theorem 2.6. *The expectation* $\mathrm{E}\xi$ *of the estimator (2.40) yields the exact solution of (2.30); i.e.* $\mathrm{E}\xi = u$; *moreover,* $\mathrm{D}\xi < C_3 < \infty$ *for all* $\mathbf{r}_0 \in \mathrm{D}$.

Proof. To prove the theorem, it suffices to use Lemma 2.4 and the results of Section 2.1. □

Expression (2.34) involves an unknown error function. For this reason, we have to examine a real biased estimator $\xi_{\varepsilon,h}$, with ϕ replaced by zero.

Theorem 2.7. *If the second-order derivatives of a solution to (2.30) are bounded in* Γ_ε, *then, under the conditions of one of the Lemmas 2.5-7, the expectation* $\mathrm{E}\xi_{\varepsilon,h} = u_{\varepsilon,h}$ *of the estimator exists; moreover,* $||\mathrm{E}\xi_{\varepsilon,h}|| < \infty$, $||\mathrm{D}\xi_{\varepsilon,h}|| < \infty$, *and* $|u(\mathbf{r}) - u_{\varepsilon,h}(\mathbf{r})| \le \mathrm{const}\, h$.

Proof. Consider the difference of the estimators

$$\xi - \xi_{\varepsilon,h} = \sum_{k=1}^{\eta} \Big(\prod_{i=0}^{j_k-1} Q(\mathbf{r}_i) \Big) \phi(\mathbf{r}_{j_k}),$$

where the sum is calculated over all points of the chain $\{\mathbf{r}_{j_k}\} \subset \{\mathbf{r}_i\}_{i=0,\dots,L}$, which lie in Γ_ε. Since, for $c \ge 0$, the weight Q does not exceed 1 and the mean number $\mathrm{E}\eta$ of hittings of Γ_ε is $O(1/h)$; recalling (2.32), we obtain:

$$\mathrm{E}|\xi - \xi_{\varepsilon,h}| \le C_4\varepsilon + C_5 h.$$

□

Now, we evalute the number $R_{\delta,n}$ of arithmetic operations needed for attaining the given error δ in the estimator for a solution at $\mathbf{r}_0$. Observe that $R_{\delta,n} \sim N_\delta Q_{\delta,n} C_n$, where N_δ is the number of modeled trajectories, $Q_{\delta,n} = \mathrm{E}L$ is the number of transitions in the chain until the termination time, $C_n \asymp n$ is the number of arithmetric operations needed for modeling a single transition in the n-dimensional space. If $N_\delta \asymp \delta^{-2}$, then, under the conditions of Lemma 2.6, for $h = \delta$, we have $R_{\delta,n} \asymp n^2\delta^{-3}$ and, under the conditions of Lemmas 2.5 and 2.7, ignoring the dependence of $\mathrm{E}\eta$ on n, we have $R_{\delta,n} \asymp n^2|\ln\delta|\delta^{-3}$.

As usual, estimating a solution to the problem, we can also estimate various functionals of a solution, partial derivatives with respect to the space variables and the parameter c, etc. Using weighted methods, we can simultaneously estimate a solution on the same trajectories at several points of the ball $\mathrm{D}(\mathbf{r}_0)$.

5. To reduce the cost of the estimator, we can consider another way of approximating the boundary conditions, providing the error order h^2 in the estimator for a solution. Suppose that the coordinate system $\{e_1, \dots, e_n\}$ is introduced in R^n and the domain D is the union of coordinate parallelepipeds. We consider the problem for the equation $\Delta u - cu = -g$ with the boundary conditions

$$\alpha u + \beta \frac{\partial u}{\partial \mathbf{n}} = \varphi, \ \mathbf{r} \in \Gamma, \tag{2.41}$$

where $\mathbf{n} = \mathbf{n}(\mathbf{r})$ is the outward normal to the boundary at $\mathbf{r} \in \Gamma$.

Using the Taylor series expansion at $\mathbf{r} \in \Gamma_\varepsilon$ along the direction of the normal $\mathbf{n}(\mathbf{b})$, we can easily obtain the equality

$$\begin{aligned}(\beta + \alpha(h + \Delta))u(\mathbf{r}) &= (\beta + \alpha\Delta)u(\mathbf{r} - h\mathbf{n}) + h\varphi(\mathbf{b}) \\ &- \left(\alpha(h^2\Delta/2 + \Delta^2 h/2) + \beta(h^2/2 + \Delta h)\right)\frac{\partial^2 u(\mathbf{r})}{\partial \mathbf{n}^2} + \phi(\mathbf{r}),\end{aligned} \tag{2.42}$$

where α and β are calculated at $\mathbf{b}$ and $\Delta = |\mathbf{r} - \mathbf{b}|$. Put $h = \varepsilon$. Assuming that the third-order partial derivatives of a solution to (2.30) are bounded in Γ_ε, we can determine the error order:

$$\max_{r \in \Gamma_\varepsilon} \left| \frac{1}{(\beta(\mathbf{r}) + \alpha(\mathbf{r})(h + \Delta))} \phi(\mathbf{r}) \right| \le C_6 \varepsilon h^2 + C_7 h^3. \tag{2.43}$$

Given $\mathbf{r} \in \Gamma_\varepsilon$, define the set $J_\mathbf{r} \subset \{1, \dots n\}$ of all indices such that, $\forall j \in J_\mathbf{r}$, the distance from $\mathbf{r}$ to Γ along the straight line with the direction vector e_j is less than h, and let J be the number of such indices. Suppose that $\{\mathbf{b}_j\}_{j \in J_\mathbf{r}} \subset \Gamma$ is the set of boundary points such that $\Delta_j = |\mathbf{b}_j - \mathbf{r}| \le h$ and $\mathbf{n}(\mathbf{b}_j) \parallel (\mathbf{b}_j - \mathbf{r})$, where $\mathbf{n}(\mathbf{b}_j) \parallel e_j$. Also define the set $\{k_j \in (0, 1] : j \in J_\mathbf{r}\}$ of coefficients such that $\sum_{j \in J_\mathbf{r}} k_j = 1$ and

$$a_0 h^2 = k_j\left(\alpha(\mathbf{b}_j)(h^2\Delta_j/2 + \Delta_j^2 h/2) + \beta(\mathbf{b}_j)(h^2/2 + \Delta_j h)\right), \ \text{for} \, \forall j \in J_\mathbf{r}.$$

Write down (2.42) for the points $\mathbf{b}_j$ and sum the results with the weights k_j:

$$\begin{aligned}\sum_{j \in J_\mathbf{r}} k_j(\beta_j + \alpha_j(h + \Delta_j))u(\mathbf{r}) &= \sum_{j \in J_\mathbf{r}} (\beta_j + \alpha_j\Delta_j)u(\mathbf{r} - h\mathbf{n}_j) \\ &+ h \sum_{j \in J_\mathbf{r}} k_j \varphi(\mathbf{b}_j) - a_0 h^2 \sum_{j \in J_\mathbf{r}} \frac{\partial^2 u(\mathbf{r})}{\partial \mathbf{n}_j^2} + \sum_{j \in J_\mathbf{r}} k_j \phi_j(\mathbf{r}),\end{aligned} \tag{2.44}$$

where $\alpha_j = \alpha(\mathbf{b}_j)$, $\beta_j = \beta(\mathbf{b}_j)$ and $\mathbf{n}_j = \mathbf{n}(\mathbf{b}_j)$. Using the difference approximation in (2.30), we arrive at the equality

$$\begin{aligned}-\sum_{j \in J_\mathbf{r}} \frac{\partial^2 u(\mathbf{r})}{\partial e_j^2} &= g(\mathbf{r}) - cu(\mathbf{r}) \\ &+ \sum_{i \notin J_\mathbf{r}} \frac{u(\mathbf{r} - he_i) + u(\mathbf{r} + he_i) - 2u(\mathbf{r})}{h^2} + O(h^2).\end{aligned} \tag{2.45}$$

Inserting (2.45) in (2.44) and finally collecting the coefficients, we obtain:

$$\begin{aligned} u(\mathbf{r}) = \sum_{j\in J_r} \frac{a_j}{s} u(\mathbf{r} - h\mathbf{n}_j) + \sum_{i\notin J_r} \frac{a_i}{s}\Big(u(\mathbf{r} - h\mathbf{e}_i) + u(\mathbf{r} + h\mathbf{e}_i)\Big) \\ + \frac{a_0 h^2}{s} g(\mathbf{r}) + \sum_{j\in J_r} \frac{k_j h}{s} \varphi(\mathbf{b}_j) + \tilde{\phi}(\mathbf{r}), \end{aligned} \tag{2.46}$$

where

$$a_j = k_j(\beta_j + \alpha_j \Delta_j), \quad j \in J_r;$$

$$a_i = a_0 = k_j(\alpha_j \Delta_j (1 + \Delta_j/h)/2 + \beta_j(1/2 + \Delta_j/h)), \quad \forall i \notin J_r;$$

$$s = \sum_{j\in J_r} a_j + 2\sum_{i\notin J_r} a_i + c a_0 h^2 + \sum_{j\in J_r} k_j \alpha_j h.$$

It is clear that, using (2.46), we can construct a new integral equation of the form (2.33) and prove an analog of Lemma 2.4 for it. Modeling the chain of a random walk on spheres with reflection from the boundary into Γ_ε, we choose the next point $\mathbf{r}_{k+1}$ as follows: $\mathbf{r}_{k+1} = \mathbf{r}_k - h\mathbf{n}_j$ with probabilities $p = \frac{a_j}{s}$, $j \in J_r$; $\mathbf{r}_{k+1} = \mathbf{r}_k \pm h\mathbf{n}_i$ with probabilities $p = \frac{a_i}{2s}$, $(i \notin J_r)$; and the trajectory terminates with probability

$$p = \frac{1}{s}\Big(\sum_{j\in J_r} k_j \alpha_j h + c a_0 h^2\Big).$$

Thus, for the new chain in the case of hitting of Γ_ε we model not only reflections in Γ but also sliding along the boundary.

Such modification reduces the number $R_{\delta,n}$ of arithmetic operations needed for attaining the given error δ in the estimator for a solution. In an analog of Theorem 2.7 stated for the new estimator, we have the error order $|\mathrm{E}\xi_{\varepsilon,h} - u(\mathbf{r})| \le Ch^2$. Thereby, under the conditions of Lemma 2.6, we have the following asymptotic order of the number of operations: $R_{\delta,n} \asymp n^2 \delta^{-2.5}$.

2.4. Solution of the Dirichlet problem for the vector and nonlinear Helmholtz equations

1. Consider the Dirichlet problem for the system of Helmholtz equations

$$\Delta u_i(\mathbf{r}) + \sum_{j=1}^{m} c_{ij}(\mathbf{r}) = -g_j(\mathbf{r}), \quad u_i(\mathbf{r})|_\Gamma = \psi_i(\mathbf{r}), \quad i = 1, \dots, m, \tag{2.47}$$

in a domain $\mathrm{D} \in R^3$ with boundary $\Gamma, \mathbf{r} = (x, y, z) \in \mathrm{D}$.

The regularity conditions are assumed to be fulfilled on the functions $\{c_{ij}, g_i, \psi_i\}$ and the boundary Γ which ensure existence and uniqueness of a solution to problem (2.47), together with the integral and probability representations (see, for instance, [20], [25]) used henceforth. Here the functions $\{c_{ij}, g_i, \psi_i\}$ may be complex-valued as well.

Denote by the symbol c^* the first eigenvalue of the scalar Helmholtz equation $\Delta u + cu = 0$ in the domain D. It is well known that if $c < c^*$ then there is a unique solution to the Dirichlet problem for the equation $\Delta u + cu = -g$, provided that g and Γ are sufficiently regular. For a nondegenerate symmetric real matrix $C = \{c_{ij}\}$ independent of $\mathbf{r}$, the condition

$$\lambda_k < c^*, \quad k = 1, ..., m, \tag{2.48}$$

represents an analog of the condition $c < c^*$; here $\{\lambda_k\}$ are the eigenvalues of the matrix C. In the case of variable matrices $\{c_{ij}(\mathbf{r})\}$, in place of (2.48) we can consider the condition

$$\sum_{j=1}^{m} |c_{ij}(\mathbf{r})| < c^*, \quad i = 1, ..., m; \quad \mathbf{r} \in \mathrm{D}.$$

We further show that under this condition a solution to problem (2.47) is unique for $\forall\{g_i, \psi_i\}$ and hence existent by the Fredholm alternative [37].

Introduce the notations: $d(\mathbf{r})$ is the distance from the point $\mathbf{r}$ to the boundary Γ;Γ_ε is the ε- neighbourhood of Γ; $S(\mathbf{r})$ and $\mathrm{D}(\mathbf{r})$ are the sphere and ball with radius $d(\mathbf{r})$ and centre the point $\mathbf{r}$; and $G_\mathbf{r}(\mathbf{r}')$ is the spherical Green function for the Laplace equation at the centre of the ball $\mathrm{D}(\mathbf{r})$, i.e.,

$$G_\mathbf{r}(\mathbf{r}') = \frac{1}{4\pi}\Big(\frac{1}{|\mathbf{r}-\mathbf{r}'|} - \frac{1}{d(\mathbf{r})}\Big), \quad |\mathbf{r}-\mathbf{r}'| \le d(\mathbf{r}).$$

Rewrite (2.47) as

$$\Delta u_i(\mathbf{r}) = -\sum_{j=1}^{m} c_{ij}(\mathbf{r})u_j(\mathbf{r}) - g_j(\mathbf{r}), \quad u_i(\mathbf{r})|_\Gamma = \psi_i(\mathbf{r}), \quad i = 1, ..., m. \tag{2.49}$$

By using the integral representation for a solution to the Poisson equation at the centre of the ball $\mathrm{D}(\mathbf{r})$, we can write down a special system of integral equations in $u_i(\mathbf{r})$. The system takes the form

$$\begin{aligned}\varphi_i(\mathbf{r}) &= \frac{1}{4\pi d^2(\mathbf{r})}\int\limits_{S(\mathbf{r})} \varphi_i(\mathbf{r}(s))ds + \int\limits_{\mathrm{D}(\mathbf{r})} G_\mathbf{r}(\mathbf{r}')\sum_{j=1}^{m} c_{ij}(\mathbf{r}')\varphi_j(\mathbf{r}')d\mathbf{r}' \\ &+ \int\limits_{\mathrm{D}(\mathbf{r})} G_\mathbf{r}(\mathbf{r}')g(\mathbf{r}')d\mathbf{r}', \quad i = 1, ..., m, \quad \mathbf{r} \in \mathrm{D}\setminus\Gamma_\varepsilon,\end{aligned} \tag{2.50}$$

in $\mathrm{D}\setminus\Gamma_\varepsilon$ and is identical in Γ_ε :

$$\varphi_i(\mathbf{r}) \equiv u_i(\mathbf{r}), \quad \mathbf{r} \in \Gamma_\varepsilon.$$

Since the standard estimator of the Monte Carlo method realizes a probability representation of Neumann series, to construct it, we must represent relation (2.50) in probabilistic form. The simplest way is to make this by analogy with [50], rewriting (2.50) as

$$\begin{aligned}\varphi_i(\mathbf{r}) &= \Big[1 - \frac{c_0 d^2(\mathbf{r})}{6}\Big]\frac{[1 - c_0 d^2(\mathbf{r})/6]^{-1}}{4\pi d^2(\mathbf{r})}\int\limits_{S(\mathbf{r})} \varphi_i(\mathbf{r}(s))ds \\ &+ \sum_{j=1}^{m}\frac{1}{m}\frac{c_0 d^2(\mathbf{r})}{6}\int\limits_{\mathrm{D}(\mathbf{r})}\frac{mc_{ij}(\mathbf{r}')}{c_0}6d^{-2}(\mathbf{r})G_\mathbf{r}(\mathbf{r}')\varphi_j(\mathbf{r}')d\mathbf{r}' \\ &+ \int\limits_{\mathrm{D}(\mathbf{r})} G_\mathbf{r}(\mathbf{r}')g_i(\mathbf{r}')d\mathbf{r}', \quad i = 1, ..., m, \quad \mathbf{r} \in \mathrm{D}\setminus\Gamma_\varepsilon.\end{aligned} \tag{2.51}$$

The positive quantity c_0 must satisfy the inequality

$$c_0 d_{\max}^2/6 < 1. \tag{2.52}$$

If we introduce one more integral-valued coordinate i in the phase space, then the system of equations (2.51) can be represented as a single integral equation in the unknown $\varphi(x) \equiv \varphi(\mathbf{r}, i) \equiv \varphi_i(\mathbf{r})$:

$$\varphi = K\varphi + h;$$

moreover, the sum over i is treated as an integral with respect to the corresponding discrete measure.

The standard estimator of the Monte Carlo method for solving an integral equation realizes a probability representation of the Neumann series and has the following form (see Section 1.3)

$$\xi_0 = \xi_{x_0} = \xi_{\mathbf{r}_0, i_0} = \sum_{k=0}^{N} Q_k h_{i_k}(\mathbf{r}_k), \quad \mathrm{E}\xi_0 \equiv \varphi(x_0) \equiv \varphi_{i_0}(\mathbf{r}_0),$$

where $\{\mathbf{r}_k\}$ is some terminating Markov chain and $\{Q_k\}$ are auxiliary weights. In accordance with the form of integral equation (2.51), such estimator can be constructed by some random walk on spheres and balls in which, with probability $1 - c_0 d^2(\mathbf{r}_k)/6$, the new point $\mathbf{r}_{k+1}$ is picked uniformly at random on the sphere $S(\mathbf{r}_k)$, the index i_{k+1} remains the same, and the weight Q_k is multiplied by the quantity

$$q(\mathbf{r}_k, c_0) = [1 - c_0 d^2(\mathbf{r}_k)/6]^{-1};$$

and with probability $c_0 d^2(\mathbf{r}_k)/6$, the index i_{k+1} equiprobably takes one of the values $j \subset (1, ..., m,)$, the point $\mathbf{r}_{k+1}$ is picked inside the ball $\mathrm{D}(\mathbf{r}_k)$ according to the density $6d^{-2}(\mathbf{r}_k)G_{\mathbf{r}_k}(\mathbf{r}')$, and the weight Q_k is multiplied by the quantity $c_{i_k, i_{k+1}}(\mathbf{r}_{k+1})m/c_0$. As a result, we obtain

$$Q_k = \Big[\prod_{i=1}^{m_k} c_{l_{k_i-1}, l_{k_i}}(\mathbf{r}_{k_i})m/c_0\Big]\Big[\prod_{i=1}^{k-m_k} q(\mathbf{r}_{t_i-1}, c_0)\Big].$$

Here $\{\mathbf{r}_{t_i}\}, i = 1, ..., k - m_k$, are points picked on spheres; $\{\mathbf{r}_{k_i}\}, i = 1, ..., m_k$, are points picked in balls; and $\{l_k\}$ is the sequence of picked indices. The functions $h_i(\mathbf{r})$ are calculated exactly or are randomized (see Section 2.1); when the point enters Γ_ε, the chain terminates and the counter increases by the estimate of a solution multiplied by the weight.

Since the exact solution in Γ_ε is unknown, instead of ξ we consider the actual estimator ξ_ε that appears when $h_i(\mathbf{r}) = \psi_i(P(\mathbf{r}))$ for $\mathbf{r} \in \Gamma_\varepsilon$, where $P(\mathbf{r})$ is a point of the boundary Γ which is nearest to $\mathbf{r}$. To justify asymptotic unbiasedness of the quantity ξ_ε as an estimator for a solution to the initial problem, we present the following assertions:

Lemma 2.8. *If the conditions*

$$c^* > m|c_{ij}(\mathbf{r})|, \quad i, j = 1, ..., m, \quad \mathbf{r} \in \mathrm{D}; \quad c_0 < 0.6079c^*$$

hold, then the Neumann series for equation (2.51) converges.

In the proof of Lemma 2.8 by a strong analogy with [50], we use a probability representation of a solution to the Dirichlet problem for the equation $\Delta u + c_0 u = C$; moreover, $\psi \equiv C$. Observe that the condition $c_0 < 0.6079c^*$ implies inequality (2.52) (see [50]).

Theorem 2.8. *Under the conditions of Lemma 2.8 , there is a unique bounded solution to equation (2.51) representable as a Neumann series and coincident with a solution to problem (2.47).*

Theorem 2.9. *If the first-order derivatives of a solution to problem (2.47) are bounded in* D, *then under the conditions of Lemma 2.8, the quantity* $\mathrm{E}\xi_\varepsilon = \varphi_\varepsilon(\mathbf{r}, i)$ *exists and*

$$|u(\mathbf{r}, i) - \varphi_\varepsilon(\mathbf{r}, i)| \le \text{const} \cdot \varepsilon, \quad \mathbf{r} \in \mathrm{D}, \quad \varepsilon > 0, \quad 1 = 1, ..., m.$$

We now turn to the question of uniform (in ε) boundedness of the variance $\mathrm{D}\xi_\varepsilon$. The following theorem holds:

Theorem 2.10. *If the conditions*

$$g_i \equiv 0, \quad c_0 > m|c_{ij}(\mathbf{r})|, \quad \mathbf{r} \in \mathrm{D}, \quad i, j = 1, ..., m; \quad c_0 < 0.488c^*$$

are satisfied, then $\mathrm{D}\xi_\varepsilon < c_d < +\infty, \varepsilon > 0$.

In the case of a nonhomogeneous system, to relax the finiteness condition for the variance, we carry out the following modification of the estimator: We shall calculate the function $h(\mathbf{r}, i)$ only in the case of the step $(\mathbf{r}, i) \to (\mathbf{r}', i)$ to a sphere, but with the weight $q(\mathbf{r}, c_0)$,i.e., $h(\mathbf{r}, i) \to h_1(\mathbf{r}, i)$:

$$h_1(\mathbf{r}_{k_i-1}, l_{k_i-1}) = 0, \quad h_1(\mathbf{r}_{t_i-1}, l_{t_i-1}) = q_1(\mathbf{r}_{t_i-1}, c_0)h(\mathbf{r}_{t_i-1}, l_{t_i-1}).$$

In place of the estimator ξ_ε, we shall thus consider the modified estimator

$$\xi_{\varepsilon,1} = \sum_{k=0}^{N} Q_k h_1(x_k) = \sum_{i=1}^{N-m_N} Q_{t_i} h(\mathbf{r}_{t_i-1}, l_{t_i-1}) + Q_N \psi(P(\mathbf{r}_N), l_N).$$

The iterated averaging yields $\mathrm{E}\xi_{\varepsilon,1} = \mathrm{E}\xi$.

Theorem 2.11. *Under the conditions of Theorem 2.10 we have*

$$\mathrm{D}\xi_{\varepsilon,1} < c_{d,1} < +\infty, \quad \varepsilon > 0, \quad \mathbf{r} \in \mathrm{D}.$$

Theorems 2.8-11 are provable by a close analogy with the corresponding assertions in [50].

We can slightly extend the scope of applicability of the constructed algorithm by simulating the integral-valued coordinate in proportion with the quantities $|c_{ij}(\mathbf{r}_{k+1})|$ rather than equiprobably; in this case Lemma 2.8 and Theorems 2.8-11 hold with the conditions

$$c^*, c_0 > \sum_j |c_{ij}(\mathbf{r})|, \quad i = 1, ..., m; \quad \mathbf{r} \in \mathrm{D} \tag{2.53}$$

substituted for the conditions $c^*, c_0 > m|c_{ij}(\mathbf{r})|$.

Alongside the algorithm, it is possible to construct a vector algorithm for solving problem (2.47). The base of the latter is system (2.50) without randomizing in j. The vector estimator for the solution $u(\mathbf{r})$ has the form

$$\zeta_0 = \sum_{k=0}^{N} Q_k H(\mathbf{r}_k), \quad u(\mathbf{r}_0) = \mathrm{E}\zeta_0,$$

where $\{\mathbf{r}_k\}$ is the chain of a random walk on spheres and balls, Q_0 is the identity matrix, and the weight Q_{k+1} is determined as follows: if $\mathbf{r}_{k+1}$ is picked on a sphere, then the matrix Q_k is multiplied by the scalar $q(\mathbf{r}_k, c_0)$; while if $\mathbf{r}_{k+1}$ is picked inside a ball, then Q_k is premultiplied by the matrix $C(\mathbf{r}_{k+1})/c_0$, where $C(\mathbf{r}) = \{c_{ij}(\mathbf{r})\}$. As a result, we obtain

$$Q_k = \Big[\prod_{i=1}^{m_k} C(\mathbf{r}_{k_i})/c_0\Big]\Big[\prod_{i=1}^{k-m_k} q(\mathbf{r}_{t_i-1}, c_0)\Big].$$

By analogy, we construct the ε-biased estimator

$$\zeta_{\varepsilon,0} = \sum_{k=0}^{N-1} Q_k H(\mathbf{r}_k) + Q_N \psi(P(\mathbf{r})),$$

where $\psi(P) = (\psi_1(P), ..., \psi_m(P))$. The assertions of Lemma 2.8 and Theorems 2.8-11 are valid for the vector estimator as well, with conditions (2.53) substituted for the conditions $c^*, c_0 > m|c_{ij}(\mathbf{r})|$.

It is easy to observe that the vector estimator is obtainable from the former scalar estimator with the help of the conditional averaging over j which diminishes the variance. The matrix operations increase the cost of vector estimation. On the other hand, the vector estimator can be especially effective in case one needs a dependent estimator for the components of a solution.

The algorithms constructed can be generalized to the case of the Dirichlet problem for an integro-differential equation of the form

$$\Delta u(\mathbf{r}, \lambda) = -\int_\Lambda u(\mathbf{r}, \lambda')c(\mathbf{r}, \lambda, \lambda')\mathrm{d}\lambda' - g(\mathbf{r}, \lambda), \quad u(\mathbf{r}, \lambda)|_\Gamma = \psi(\mathbf{r}, \lambda).$$

In this case, if during a random walk on spheres and balls the new point $\mathbf{r}_{k+1}$ is picked inside a ball, then the new value λ_{k+1} is picked according to the density $p(\mathbf{r}_{k+1}, \lambda_k, \lambda')$ and the weight is multiplied by the quantity $c(\mathbf{r}_{k+1}, \lambda_k, \lambda_{k+1})/(c_0 p(\mathbf{r}_{k+1}, \lambda_k, \lambda_{k+1}))$. The assertions of Lemma 2.8 and Theorems 2.8-11 remain valid, provided that the conditions $c^*, c_0 > m|c_{ij}(\mathbf{r})|$ are replaced with the conditions

$$c^*, c_0 > |c(\mathbf{r}, \lambda, \lambda')|/p(\mathbf{r}, \lambda, \lambda'), \quad \mathbf{r} \in \mathrm{D}, \quad \lambda, \lambda' \in \Lambda.$$

By setting

$$p(\mathbf{r}, \lambda, \lambda') = \frac{|c(\mathbf{r}, \lambda, \lambda')|}{\int_\Lambda |c(\mathbf{r}, \lambda, \lambda')| d\lambda'},$$

we obtain the following condition for existence and uniqueness of a solution to the integro-differential equation under study:

$$\int_\Lambda |c(\mathbf{r}, \lambda, \lambda')|\mathrm{d}\lambda' < c^*, \quad \mathbf{r} \in \mathrm{D}, \quad \lambda \in \Lambda.$$

It is easy to see that in the case of piecewise constant functions $c_{ij}(\mathbf{r})$, the estimators depend polynomially on the values of the functions, which allows us to construct easily the ε-biased estimators for the parametric derivatives [50]. Observe that the estimators for the parametric derivatives can effectively be used for the iterative solution of inverse problems in parameters with appropriate regularization (see, for instance, [41]).
2. Let us consider the following boundary-value problem in a bounded domain $D \subset R_3$ with a piecewise smooth boundary Γ :

$$\Delta u + u^n = 0, \quad u|_\Gamma = \psi, \quad n \geq 2, \tag{2.54}$$

where $\psi \in C(\Gamma)$. The function u is considered in $C_2(D) \cap C(\bar{D})$ [62]. Along with (2.54) we consider the parametrized problem

$$\Delta u_c + u_c^n = 0, \quad u_c|_\Gamma = c\psi, \quad c \geq 0, \tag{2.55}$$

On the strength of the results in [62] one can argue that there exists $c^{(0)} > 0$ such that for $c \leq c^{(0)}$ a solution of problem (2.55) exists and is unique in the ball:

$$n\|u_c^{n-1}\| \leq c^*, \quad \|u_c^{n-1}\| = \max_{\mathbf{r}\in D} |u_c^{n-1}(\mathbf{r})|,$$

where $c^* > 0$ is the first eigenvalue of the problem

$$\Delta u + cu = 0, \quad u|_\Gamma = 0.$$

As in Section 1.6, here the uniqueness can easily be established by considering the difference between $u_{c,1}$ and $u_{c,2}$ which are solutions of problem (2.55). If $\psi \geq 0$, then the monotonic increase of u_c and therefore $\|u_c\|$ with c is established as in Section 1.6 by means of the iterative transition from u_{c_2} to u_{c_1} $[c_1 < c_2 < c^{(0)}]$ on the basis of an integral equation [62] that is equivalent to problem (2.55) and is determined by the Green function for D.

For sufficiently small values of c, the parametric derivatives $u^{(k)}$ are the solutions of the following boundary-value problems:

$$\Delta u' + nu^{n-1}u' = 0, \quad u'|_\Gamma = \psi, \quad \Delta u^{(k)} + (u^n)^{(k)} = 0, \quad u^{(k)}|_\Gamma = 0, \quad k > 1. \tag{2.56}$$

Some conditions for the validity of relations (2.56) will be established below through the analysis of the corresponding probabilistic representations.

In present section we construct Monte Carlo estimates in the random walk on spheres and balls (see Item 1) for the problems involved. Branching occurs when the trajectory hits a ball. Branching chains of this kind with additional absorption were considered in [20] in order to estimate the solution of the equation $\Delta u = u^2$. New estimates presented here are characterized by pure multiplication. They call for special investigation, using the result of Item 1.

Let us introduce the notation: $d(\mathbf{r})$ is a distance between the point $\mathbf{r}$ and the boundary Γ, Γ_ε is the ε-vicinity of the boundary, $S(\mathbf{r})$ is the sphere of radius $d(\mathbf{r})$ with centre at the point $\mathbf{r}$, $D(\mathbf{r})$ is the corresponding ball.

By analogy with Item 1 we consider the following integral representation of the solution:

$$u_1(\mathbf{r}) = \frac{1}{4\pi d^2(\mathbf{r})} \int_{S(\mathbf{r})} u_1(\mathbf{r}(s))ds + \int_{D(\mathbf{r})} u_1^n(\mathbf{r}')G_\mathbf{r}(\mathbf{r}')d\mathbf{r}', \quad \mathbf{r} \in D\backslash\Gamma_\varepsilon. \tag{2.57}$$

For $\mathbf{r} \in \Gamma_\varepsilon$ one uses the identical representation $u_1(\mathbf{r}) \equiv u(\mathbf{r})$. Its approximation by the boundary condition ($u \approx \tilde{u}$) leads to the ε-biased estimate (see Section 2.1). Here $G_\mathbf{r}(\mathbf{r}')$ is the Green function for a ball with the pole at the centre of the ball:

$$G_{\mathbf{r}}(\mathbf{r}') = \frac{1}{4\pi}\Big[\frac{1}{|\mathbf{r}-\mathbf{r}'|} - \frac{1}{d(\mathbf{r})}\Big], \quad |\mathbf{r}-\mathbf{r}'| \le d(\mathbf{r}).$$

In order to construct the Monte Carlo estimate for problem (2.54), by analogy with Item 1, one realizes a random walk on spheres and balls from the point $\mathbf{r}_0 = \mathbf{r}$. In so doing, a new point $\mathbf{r}_{i+1}$ is chosen uniformly on the sphere $S(\mathbf{r}_i)$ with the probability $1 - c_0 d^2(\mathbf{r}_i)/6$, and the weight Q is multiplied by

$$q(\mathbf{r}_i, c_0) = [1 - c_0 d^2(\mathbf{r}_i)/6]^{-1};$$

the point $\mathbf{r}_{i+1}$ is chosen within the ball $D(\mathbf{r}_i)$ with the complementary probability $c_0 d^2(\mathbf{r}_i)/6$ according to the density $6d^{-2}(\mathbf{r}_i)G_{\mathbf{r}_i}(\mathbf{r}')$ and the weight is multiplied by c_0^{-1}. In the last case, n-multiple branching is performed, i.e. n conditionally-independent random walk trajectories are constructed from the point $\mathbf{r}_i$. Here $c_0 < 0.6079c^*$ (see Item 1). When Γ_ε is hit, the corresponding branch of a trajectory terminates and the quantity Qu (or $Q\tilde{u}$ if one goes to the real ε-biased estimate) is added to the value of the estimate. Further (see Theorem 2.13), we will show that under certain conditions the branching chain involved terminates with probability 1. The constructed estimate is of the following recurrent form:

$$\xi_0 = \begin{cases} [1 - c_0 d^2(\mathbf{r})/6]^{-1}\xi_1, & P = 1 - c_0 d^2(\mathbf{r})/6 \\ \Big(\prod_{k=1}^{n} \xi_1^{(k)}\Big)/c_0, & P = c_0 d^2(\mathbf{r})/6. \end{cases}$$

Continuing the recursion up to the end of the branching trajectory, we find the following probabilistic representation of the Neumann series for problem (2.54) (such constructions are fully considered in detail, for example, in [20]). The next theorem assumes the branching chain to terminate with probability 1 and $\varepsilon = 0$.

Theorem 2.12. *If $c^{(0)} \ge 1$ and $c_0 < 0.6079c^*$ then $\mathrm{E}\xi \equiv u$, with $n\|u^{n-1}\| < c^*$.*

Proof. At first we assume that $\psi \ge 0$. Then any rearrangements of summations and preliminary conditional averaging procedures are allowed in the process of averaging, since all the elements of the estimate ξ are non-negative. Thus, $\mathrm{E}\xi = \mathrm{E}\zeta$ and

$$\zeta_0 = \begin{cases} [1 - c_0 d^2(\mathbf{r})/6]^{-1}\zeta_1, & P = 1 - c_0 d^2(\mathbf{r})/6 \\ (\mathrm{E}\xi_1)^{n-1}\zeta_1/c_0, & P = c_0 d^2(\mathbf{r})/6. \end{cases}$$

Therefore, if the inequality

$$\|(\mathrm{E}\xi)^{n-1}\| < c^*/n$$

is valid, then taking into account the inequality $c_0 < 0.6079c^*$, one can state on the basis of Item 1 that the function $\mathrm{E}\xi$ is the solution of problem (2.54). Let us now assume

$$\|(\mathrm{E}\xi)^{n-1}\| \ge c^*/n. \tag{2.58}$$

As in Section 1.6, here we arrive at a monotonic and continuous (with respect to c) increase of the function $\|\mathrm{E}\xi(c)\|$ up to $+\infty$, where $\xi(c)$ is the estimate of the solution of the parametrized problem. Hence if inequality (2.58) and the inequality $c^{(0)} > 1$ are valid, one can find such $c < 1$ that the following relation holds:

$$\|u^{n-1}\| < \|(\mathrm{E}\xi(c))^{n-1}\| < c^*/n.$$

However, in the case $\mathrm{E}\xi(c) \equiv u_c$ and $\|u_c^{n-1}\| \leq \|u^{n-1}\|$, since $c < 1$. The contradiction obtained proves the theorem for $\psi \geq 0$. Going to the case with alternating signs, the argument is performed in the same manner as in Section 1.6, by using the estimate ξ for problem (2.55) as the majorant with the substitution $|\psi| \to \psi$. Theorem 2.12 is proved. □

Below we give a sufficient condition for the finiteness of the average number of branches that is necessary in order to make calculations possible.

Theorem 2.13. *If $c_0 < 0.6079c^*$ and $nc_0 < c^*$, then the average number of branches that have reached the boundary in the random walk under consideration for $\varepsilon = 0$ is majorized with the solution of the problem $\Delta u + nu = 0$, $u|_\Gamma = 1$.*

The proof follows immediately from Item 1 with allowance made for the fact that branching is, on average, equivalent to multiplying the auxiliary weight by n. □

Theorem 2.13 gives a logarithmic estimate of the complexity: $S_\varepsilon \asymp |\ln\varepsilon|$, with the known suppositions of the regularity of the boundary Γ [20]. Note that with the uniform boundedness of the derivatives of the solution with respect to coordinates in D, when $\mathbf{r} \in \Gamma_\varepsilon$, the substitution of $u(\mathbf{r})$ for the value of ψ at the point of the boundary nearest to $\mathbf{r}$ leads to the ε-biased estimate (see Item 1). In view of Theorem 2.13 we will further assume that $c_0 > 1.56$ (i.e. $c^*/n > 1.56$).

Let us now consider the quantity $\mathrm{E}\xi^2$. In this connection we need to consider Monte Carlo estimates for the problem

$$\Delta u_1 + u_1\Big(u_1^{n-1} + \frac{1}{2}c^*\Big) = 0, \quad u_1|_\Gamma = \psi. \tag{2.59}$$

By analogy with problem (2.54), on the basis of the results in [62] we assume that for $c \leq c_1^{(0)}$, the solution $u_{1,c}$ of the parametrized problem (2.59) exists and is unique in the ball

$$2n\|u_{1,c}^{n-1}\| < c^*$$

and also monotonicaly increases with c.

Theorem 2.14. *If $c_1^{(0)} \geq 1$ and $1.56 < c_0 < 0.39c^*$, then $\|\mathrm{E}\xi^2\| < \|u_1\|$, with $\|u_1^{n-1}\| < c^*/(2n)$.*

Proof. See in [50], Section 6.2. □

The estimates of the parametric derivatives $u^{(k)}$ for $\varphi \geq 0$ can be obtained by the direct differentiation of the estimate ξ (the argument in the case with alternating signs can be made as before by using majorant derivatives). As this takes place, one obtains randomized estimates of the solutions to Helmholtz equations (2.56). On this basis one can obtain sufficient conditions for finiteness of such derivatives. For example, for u' we have the following probabilistic representation:

$$\xi' = \begin{cases} [1 - c_0 d^2(\mathbf{r})/6]^{-1}\xi', & P = 1 - c_0 d^2(\mathbf{r})/6 \\ n\xi_1' \left(\prod_{k=1}^{n-1} \xi_1^{(k)} \right) / c_0, & P = c_0 d^2(\mathbf{r})/6 \end{cases}$$

and the value of $\|\mathrm{E}\xi'\|$ is obviously finite, if $n\|u^{n-1}\| < c_0 < 0.6079c^*$. We also find from Item 1 that $\|\mathrm{E}\xi'^2\| < +\infty$ if

$$n\|(\mathrm{E}\xi^2)^{(n-1)/2}\| < c_0 < 0.488c^*.$$

Note that for $c = 0$ from (2.56) there follows a sequence of Poisson equations that can be solved using standard randomized estimates with a finite variance.

The results obtained can be extended to equations of the form:

$$\Delta u_1 + u_1 \sum_{k=0}^{n} a_k u^k = 0, \quad u|_\Gamma = \psi > 0,$$

and to the similar systems of equations, for instance, to the system:

$$\Delta u_1 + a_1 u_1 u_2 = 0, \quad \Delta u_2 + a_2 u_1 u_2 = 0.$$

To extend these results to nonhomogeneous equations, for instance, to the equation $\Delta u + u^n = -g$, it is necessary to use the parametrization of the type:

$$\Delta u_c + u_c^n = -cg, \quad u_c|_\Gamma = c\psi.$$

Here, only the method of investigating the variance changes essentially.

The results obtained can be interpreted in the following way, without regard to the properties of the solution of parametrized problem (2.54) which were formulated at the beginning of the Item. The continuous (in c) function $u_c = \mathrm{E}\xi$ (for $\varepsilon = 0$ and $\tilde{u} \equiv c\psi$), which is the solution of problem (2.54) with the substitution $\psi \to c\psi$, is found, provided that $\|u_c^{n-1}\| < c^*$ and $n < 0.6079c^*$. The equality $u \equiv 0$ is valid for this function at $c = 0$. The parametric derivative $u^{(k)}$ is the solution of the corresponding Dirichlet problem (2.56) under the additional constraints of the form $n\|u^{n-1}\| < c^*$ for $k = 1$.

3. Now let us construct randomized estimates for the first and the second partial coordinate derivatives from the solution u of problem (2.54)[39].

Denote $f(u) = u^n$ and consider the integral representation [11]:

$$u(\mathbf{r}) = -\int_{S_0} \frac{\partial G(\mathbf{r}, \mathbf{r}')}{\partial n_{\mathbf{r}'}} u(\mathbf{r}') dS_{\mathbf{r}'} + \int_{D_0} G(\mathbf{r}, \mathbf{r}') f(u(\mathbf{r}')) d\mathbf{r}', \quad \mathbf{r} \in D_0, \tag{2.60}$$

where $D_0 = D(\mathbf{r}_0)$ is the ball with the centre $\mathbf{r}_0 \in D$ and the radius $\alpha d_0, 0 < \alpha \le 1$, and

$$S_0 = \partial D_0, \qquad d_0 = d(\mathbf{r}_0) = \inf|\mathbf{r} - \mathbf{r}_0|, \qquad \mathbf{r} \in \partial D,$$

$$\mathbf{r}_0 = \left(x_1^{(0)}, x_2^{(0)}, x_3^{(0)}\right), \quad \mathbf{r} = (x_1, x_2, x_3), \quad \mathbf{r}' = (x_1', x_2', x_3'),$$

$$\begin{aligned} G(\mathbf{r},\mathbf{r}') &= \frac{1}{4\pi|\mathbf{r}-\mathbf{r}'|} - \frac{d_0|\mathbf{r}'-\mathbf{r}_0|}{4\pi\left||\mathbf{r}'-\mathbf{r}_0|^2(\mathbf{r}-\mathbf{r}_0)-d_0^2(\mathbf{r}'-\mathbf{r}_0)\right|}, \\ \frac{\partial G(\mathbf{r},\mathbf{r}')}{\partial n_{\mathbf{r}'}} &= \frac{|\mathbf{r}-\mathbf{r}_0|^2-d_0^2}{4\pi d_0|\mathbf{r}-\mathbf{r}'|^3}, \qquad \mathbf{r}'\in S_0. \end{aligned} \tag{2.61}$$

After differentiating (2.60) with respect to x_i at the point $\mathbf{r}$, we have

$$\begin{aligned} \frac{\partial u(\mathbf{r})}{\partial x_i} &= -\int_{S_0} \frac{\partial}{\partial x_i}\left(\frac{\partial G(\mathbf{r},\mathbf{r}')}{\partial \mathbf{n}_{\mathbf{r}'}}\right) u(\mathbf{r}')\mathrm{d}S_{\mathbf{r}'} \\ &+ \int_{D_0} \frac{\partial}{\partial x_i}\left(G(\mathbf{r},\mathbf{r}')\right) f\left(u(\mathbf{r}')\right)\mathrm{d}\mathbf{r}', \end{aligned} \tag{2.62}$$

where

$$\begin{aligned} \frac{\partial}{\partial x_i}\left(\frac{\partial G(\mathbf{r},\mathbf{r}')}{\partial \mathbf{n}_{\mathbf{r}'}}\right) &= \frac{2\left(x_i - x_i^{(0)}\right)}{4\pi d_0|\mathbf{r}-\mathbf{r}'|^3} - \frac{3\left(d_0^2-|\mathbf{r}-\mathbf{r}_0|^2\right)(x_i'-x_i)}{4\pi d_0|\mathbf{r}-\mathbf{r}'|^5}, \\ \frac{\partial}{\partial x_i}\left(G(\mathbf{r},\mathbf{r}')\right) &= \frac{x_i'-x_i}{4\pi|\mathbf{r}-\mathbf{r}'|^3} \\ &+ \frac{d_0|\mathbf{r}'-\mathbf{r}_0|^3\left(|\mathbf{r}'-\mathbf{r}_0|^2(x_i-x_i^{(0)}) - d_0^2(x_i'-x_i^{(0)})\right)}{4\pi\left||\mathbf{r}'-\mathbf{r}_0|^2(\mathbf{r}-\mathbf{r}_0)-d_0^2(\mathbf{r}'-\mathbf{r}_0)\right|^3}. \end{aligned} \tag{2.63}$$

If $\mathbf{r}=\mathbf{r}_0$, then (2.62) is transformed to

$$\frac{\partial u(\mathbf{r}_0)}{\partial x_i} = \int_{S_0} \frac{3a_i}{4\pi d_0^3} u(\mathbf{r}')\mathrm{d}S_{\mathbf{r}'} + \int_{D_0} \frac{a_i\left(d_0^3-|\mathbf{r}'-\mathbf{r}_0|^3\right)}{4\pi d_0^3|\mathbf{r}'-\mathbf{r}_0|^2} f(u(\mathbf{r}'))\mathrm{d}\mathbf{r}', \tag{2.64}$$

where $a_i = \cos(\widehat{\mathbf{r}'-\mathbf{r}_0, e_i}), i=1,2,3$, are the cosines between the vector $(\mathbf{r}'-\mathbf{r}_0)$ and the coordinate orths (e_1, e_2, e_3).

In accordance with (2.62), the random walk runs out from the point $\mathbf{r}_0$, and with the probability $1-3d_0c_1/4$, a new point $\mathbf{r}_1$ is uniformly chosen on the sphere $S(\mathbf{r}_0)$. For this point the solution is estimated with the weight $3[d_0(1-3d_0c_1/4)]^{-1}a_i$. With the complementary probability $3d_0c_1/4$, the point $\mathbf{r}_1$ is chosen within the ball $D(\mathbf{r}_0)$ according to the density

$$\frac{1}{4\pi}\frac{4(d_0^3-\rho^3)}{3d_0^4\rho^2}, \qquad \rho = |\mathbf{r}_1-\mathbf{r}_0|.$$

For the chosen point, n conditionally independent estimates $\{\xi_1^{(k)}\}$ of the solution are constructed and multiplied by the weight a_i/c_1. Here the additional condition $0 < c_1 < 4/(3d_0)$ is required. The necessary estimate is as follows:

$$\eta_i = \begin{cases} \dfrac{3a_i}{d_0(1-3d_0c_1/4)}\xi_1, & P = 1-3d_0c_1/4; \\ \dfrac{1}{c_1}a_i\prod\limits_{k=1}^{n}\left(\xi_1^{(k)}\right), & P = 3d_0c_1/4. \end{cases} \tag{2.65}$$

Theorem 2.15. *1) Under the conditions of Theorem 2.12, the following relations hold:*

$$\mathrm{E}\eta_i = \frac{\partial u}{\partial x_i}, \quad i = 1, 2, 3.$$

2) Under the conditions of Theorem 2.14, the relations $\mathrm{D}\eta_i < +\infty$, $i = 1, 2, 3$, *hold.*

Proof. The substitution $|\psi| \to \psi$ gives us the estimates ζ_i, so that $|\eta_i| \le \zeta_i$ and $\eta_i^2 \le \zeta_i^2$. Further, the proof is performed on the basis of the Fubini theorem using Theorem 2.12 and 2.14. □

To construct estimates for the second partial derivatives we use the representation

$$G(\mathbf{r}, \mathbf{r}') = \frac{1}{4\pi|\mathbf{r} - \mathbf{r}'|} + g(\mathbf{r}, \mathbf{r}')$$

and the Green-Ostrogradsky formula

$$\begin{aligned}\frac{\partial u(\mathbf{r})}{\partial x_i} &= -\int\limits_{S_0} \frac{\partial}{\partial x_i}\left(\frac{\partial G(\mathbf{r}, \mathbf{r}')}{\partial \mathbf{n}_{\mathbf{r}'}}\right) u(\mathbf{r}')\mathrm{d}S_{\mathbf{r}'} \\ &- \int\limits_{S_0} \left(\frac{1}{4\pi|\mathbf{r} - \mathbf{r}'|}\right) f(u(\mathbf{r}')) \cos(\widehat{\mathbf{n}_{\mathbf{r}'}, e_i})\mathrm{d}S_{\mathbf{r}'} \\ &+ \int\limits_{D_0} \left(\frac{1}{4\pi|\mathbf{r} - \mathbf{r}'|}\right) \frac{\partial f(u(\mathbf{r}'))}{\partial x_i'}\mathrm{d}\mathbf{r}' + \int\limits_{D_0} \frac{\partial g(\mathbf{r}, \mathbf{r}')}{\partial x_i} f(u(\mathbf{r}'))\mathrm{d}\mathbf{r}'.\end{aligned} \tag{2.66}$$

It is possible to differentiate (2.66) under the integral signs. As a result, we have

$$\begin{aligned}\frac{\partial^2 u(\mathbf{r}_0)}{\partial x_j \partial x_i} &= \int\limits_{S_0} \left[\frac{5}{d_0^2} a_{ij} u(\mathbf{r}') - a_i a_j u^n(\mathbf{r}')\right] \frac{\mathrm{d}S_{\mathbf{r}'}}{4\pi d_0^2} \\ &+ \int\limits_{D_0} \left[n d_0 a_j \frac{\partial u(\mathbf{r}')}{\partial x_i'} - \frac{r^4}{d_0^4} a_{ij} u(\mathbf{r}')\right] u^{n-1}(\mathbf{r}') \frac{\mathrm{d}\mathbf{r}'}{4\pi d_0 r^2}.\end{aligned}$$

where $a_{ij} = 3a_i a_j - \delta_{ij}$. On this basis the following estimate is constructed:

$$\zeta_{ij} = \begin{cases} \dfrac{1}{1-p}\left(\dfrac{5}{d_0^2} a_{ij} \xi_1^{(1)} - a_i a_j \prod\limits_{k=1}^{n} \xi_1^{(k)}\right), & P = 1 - p; \\ \dfrac{1}{p}\left(n d_0 a_j \eta_{1,i} - \dfrac{\rho^4}{d_0^4} a_{ij} \xi_1^{(1)}\right) \prod\limits_{k=2}^{n} \xi_1^{(k)}, & P = p, \end{cases}$$

where $\rho = |\mathbf{r}_1 - \mathbf{r}_0|, a_{ij} = 3a_i a_j - \delta_{ij}$ and δ_{ij} are the Kronecker symbols. Here with the probability p, a new point $\mathbf{r}_1$ is uniformly chosen within the ball D_0, and with the complementary probability $(1 - p)$, on the sphere S_0. For the point $\mathbf{r}_1, n$ conditionally independent estimates $\{\xi_1^{(k)}\}$ of the solution are constructed; for a 'ball' point, in addition, the estimate $\eta_{1,i}$ of the first derivative $\partial u(\mathbf{r}_1)/\partial x_i$ is computed, here $i = 1, 2, 3$.

Theorem 2.16. *1) Under the conditions of Theorem 2.12 and* $\alpha < 1$, *the following relations hold:*

$$\mathrm{E}\zeta_{ij} = \frac{\partial^2 u}{\partial x_i \partial x_j}, \quad i,j = 1,2,3.$$

2) Under the conditions of Theorem 2.14 and $\alpha < 1$, *the relations* $\mathrm{D}\xi_{ij} < +\infty$ *hold.*

Proof. Similar to the proof of Theorem 2.15. □

To optimize the above algorithms it is possible to choose the proper values of the parameters α and p.

2.5. Estimating the main eigenvalue of the Laplace operator

1. Let us consider in $L_\infty(X)$ the integral equation dependent on the parameter:

$$\varphi(x,\lambda') = \int_X k(x,y,\lambda')\varphi(y,\lambda')\mathrm{d}y + h(x,\lambda'), \tag{2.67}$$

or, in the operator form: $\varphi = K\varphi + h$. Here X is the n-dimensional Euclidean space, $\lambda' \in \Lambda, \Lambda = [\lambda - \varepsilon_0, \lambda + \varepsilon_0]$. It is necessary to compute the derivatives

$$\varphi^{(i)}(x,\lambda) = \frac{\partial^i \varphi(x,\lambda)}{\partial \lambda^i}, \quad i = 1,\dots,m.$$

By $K^{(i)}$ we denote the integral operator with the kernel $k^{(i)}(x,y,\lambda)$, and by $\rho(K)$, the spectral radii of the operator K. It is supposed that for $\lambda' \in \Lambda$ and $i = 0,\dots,m$ the following relations hold:

$$\begin{gathered}\rho(K) < 1, \;\; k^{(i)}(\cdot,\cdot,\lambda') \geq 0, \;\; \|K^{(i)}\| < +\infty, \\ \|h^{(i)}(x,\lambda')\| \leq \tilde{h}(x), \;\; \tilde{h}(x) \in L_\infty, \;\; \|h^{(i)}\| < +\infty.\end{gathered} \tag{2.68}$$

To construct the Monte Carlo estimates, the stationary Markov chain $\{x_n\}$ $(n = 0,1,\dots,N)$ with the transition density $p(x,x')$ is considered. Here the value

$$p(x) = 1 - \int_X p(x,x')\mathrm{d}x' \geq 0$$

is considered as the probability of termination at the transition $x \to x'$; N is a random number of the latter state. It is assumed that $p(x,x') \neq 0$, if $k(x,x',\lambda') \neq 0$ at least for one value $\lambda' \in \Lambda$. Further, the weights $\{Q_n\}$:

$$Q_0 \equiv 1, \;\; Q_n(\lambda) = Q_{n-1}(\lambda)\frac{k(x_{n-1},x_n,\lambda)}{p(x_{n-1},x_n,\lambda)}$$

and the collision estimate:

$$\xi_x(\lambda) = \sum_{n=0}^{N} Q_n(\lambda) h(x_n, \lambda), \quad x_0 = x, \tag{2.69}$$

are introduced. Under the above conditions, $E\xi_x(\lambda) = \varphi(x, \lambda)$. In addition (see Section 1.4), $E\xi_x^2(\lambda) < +\infty$ (i.e., $D\xi_x(\lambda) < +\infty$), if $\rho(K_p) < 1$, where K_p is the integral operator with the kernel $k^2(\cdot,\cdot,\lambda)/p(\cdot,\cdot)$.

Using Theorem 1.14 we state that under the conditions (2.68) and $\rho(K_p) < 1$, the following relations

$$\varphi^{(i)}(x, \lambda) = E\xi_x^{(i)}(\lambda) = \sum_{n=0}^{N} [Q_n(\lambda) h(x_n, \lambda)]^{(i)}, \quad D\xi_x^{(i)}(\lambda) < +\infty$$

hold, if, additionally, the operators with the kernels $[k^{(i)}(\cdot,\cdot,\lambda)]^2/p(\cdot,\cdot)]$ are bounded, here $i = 0, 1, ..., m$.

2. Consider the following three-dimensional problem in a domain $D \in R^3$ with the boundary Γ :

$$\Delta u - cu = -g, \quad \mathbf{r} \in D; \quad \alpha u + \beta \frac{\partial u}{\partial l} = \varphi, \quad \mathbf{r} \in \Gamma, \tag{2.70}$$

where

$$0 \le \alpha(\mathbf{r}) < +\infty, \quad 0 \le \beta(\mathbf{r}) < +\infty, \quad c \equiv \text{const} > 0, \quad \alpha + \beta \ge 1.$$

The vector field $\mathbf{l}$ on Γ satisfies the following conditions:

$$(\mathbf{l}(\mathbf{r}), \mathbf{n}(\mathbf{r})) > \theta_0 > 0, \quad \|\mathbf{l}\| = 1, \quad \mathbf{r} \in \Gamma,$$

where $\mathbf{n}(\mathbf{r})$ is the outward normal vector to Γ at $\mathbf{r}$. The corresponding unbiased Monte Carlo estimate for $u(\mathbf{r})$ is determined in Section 2.4 by expression (2.40).

Here we use the jump lenght (see Section 2.4) $h = C_h \varepsilon^{1/2}$, where the constant C_h is chosen so that $\mathbf{r}_{k+1} \in D \setminus \Gamma_\varepsilon$, if $\mathbf{r}_k \in \Gamma_\varepsilon$. However, expression (2.40) involves an unknown error function ψ; to pass to the real biased estimate ξ_ε, it is necessary to replace ψ by zero.

Further, we assume that $p(\mathbf{r}) = 1 - s(c_0, \mathbf{r})$ for $\mathbf{r} \in D \setminus \Gamma_\varepsilon$; here $0 < c_0 \le c$ (c_0 may be dependent on $\mathbf{r}$), and $p(\mathbf{r}) = 1 - q(\mathbf{r})$ for $\mathbf{r} \in \Gamma_\varepsilon$.

Denote: $E\xi_\varepsilon = u_\varepsilon$. In Section 2.4, the following relations are abtained

$$|u(\mathbf{r}) - u_\varepsilon(\mathbf{r})| \le C_0 \varepsilon^{\frac{1}{2}}, \quad D\xi_\varepsilon \le C_d < +\infty, \quad \forall \varepsilon > 0.$$

provided that the second-order derivatives of $u(\mathbf{r}), \mathbf{r} \in \Gamma_\varepsilon$, are uniformly bounded.

Let us include the parameter c into the denotions of the solution and the collision estimate: $u(\mathbf{r}) = u(\mathbf{r}, c), \xi_r = \xi_r(c)$, and study the possibility of estimating the parametric derivatives $u^{(m)}(\cdot, c)$ with the use of $\xi_r^{(m)}(c)$. Further, we denote: $\lambda = -c$ and $s_0(\lambda, d) = s(-\lambda, d)$.

Lemma 2.9. *For $\lambda \le 0$, the relation $0 \le s_0^{(m)}(\lambda, d) \le C_d d^2$ holds.*

Proof. The value $s_0(\lambda, d)$ represents the solution of the following Dirichlet problem:

$$\Delta u + \lambda u = 0, \quad \lambda \le 0, \quad u|_{S(\mathbf{r})} = 1, \tag{2.71}$$

in the centre of the sphere $S(\mathbf{r})$, so that $d(\mathbf{r}) = d$. Multi-differentiation of problem (2.71) gives us the following problem:

$$\Delta u^{(m)} + \lambda u^{(m)} = -mu^{(m-1)}, \; u^{(m)}|_\Gamma = 0, \; m = 1, 2, \ldots \tag{2.72}$$

The integral representation shows that the solution of the problem (2.72) is positive, i.e. really $s^{(m)}(\lambda, d) \geq 0$. Hence $s_0^{(m)}(\lambda, d) = \left(\frac{\sqrt{t}}{sh\sqrt{t}}\right)^{(m)}_{t=d^2\lambda} d^{2m} \leq C(m)\left(\frac{\sqrt{t}}{sh\sqrt{t}}\right)^{(m)}_{t=0} d^2$, $m = 1, 2, \ldots$, because $d \leq d_{\max} < +\infty$. □

Lemma 2.10. *For the collision estimate $\xi_r = \xi_r(\lambda)$ under the conditions $\varphi \equiv 1, \lambda < 0$ and $c_0 \leq c$, the following relations hold:*

$$\mathrm{E}\xi_r^{(m)}(\lambda) = u^{(m)}(\mathbf{r}, \lambda), \quad \mathrm{E}[\xi_r^{(m)}(\lambda)]^2 < +\infty.$$

Proof. For the corresponding integral equation (see Section 2.4) we have $\rho(K) < 1$. Since for $\mathbf{r}' \in D \setminus \Gamma_\varepsilon$ the equality $p(\mathbf{r}') = 1 - s(c_0, d')$ holds; therefore

$$\frac{k^2(\mathbf{r}', \mathbf{r}, c)}{p(\mathbf{r}', \mathbf{r})} = \frac{s(c, d')}{s(c_0, d')} k(\mathbf{r}', \mathbf{r}, c) \leq k(\mathbf{r}', \mathbf{r}, c)$$

for $\mathbf{r}' \in D \setminus \Gamma_\varepsilon$, and $k^2(\mathbf{r}', \mathbf{r}, c)/p(\mathbf{r}', \mathbf{r}) = k(\mathbf{r}', \mathbf{r}, c)$ for $\mathbf{r}' \in \Gamma_\varepsilon$. Taking into account Lemma 2.9 we state that here the conditions of the Theorem are valid. □

Theorem 2.17. *Assume that for problem (2.70) with $\varphi \equiv 0$, the uniform boundedness of the second derivatives of the function g in the domain D yields the same property of the solution u. Then if $g \equiv 0$, $\varphi \equiv 1$ and $\lambda < 0$ the following inequality holds:*

$$|u^{(m)}(\mathbf{r}) - \mathrm{E}\xi_{r,\varepsilon}^{(m)}| \leq C_m \varepsilon^{\frac{1}{2}}, \; \mathbf{r} \in D, \; m = 1, 2, \ldots$$

Proof. The m-fold parametrical differentiation of relation (2.31) gives:

$$u^{(m)}(\mathbf{r}) = \frac{\beta + \alpha\Delta}{\beta + \alpha\Delta + \alpha h} u^{(m)}(\mathbf{r} - h\mathbf{l}) + \psi^{(m)}(\mathbf{r}), \; \mathbf{r} \in \Gamma_\varepsilon.$$

On the other hand, parametrical differentiation of the initial problem yields the relation

$$\Delta u^{(m)} + \lambda u^{(m)} = -mu^{(m-1)}; \; \alpha u^{(m)} + \beta \frac{\partial u^{(m)}}{\partial l^m} = 0, \; \mathbf{r} \in \Gamma_\varepsilon. \tag{2.73}$$

So, under the Theorem assumption, we have $|\psi^{(m)}| \leq C_m \varepsilon$, $m = 0, 1, \ldots$. For the estimation ξ we can write down

$$\xi_r - \xi_{r,\varepsilon} = \sum_{\mathbf{r}_i \in \Gamma_\varepsilon} \Big(\prod_{j=0}^{i-1} Q_j \Big) \psi(\mathbf{r}_i). \tag{2.74}$$

Differentiating this relation, we obtain

$$\xi_r^{(m)} - \xi_{r,\varepsilon}^{(m)} = \sum_{k=0}^{m} \sum_{\mathbf{r}_i \in \Gamma_\varepsilon} C_m^k \Big(\prod_{j=0}^{i-1} Q_j \Big)^{(k)} \psi^{(m-k)}(\mathbf{r}_i).$$

Taking into account an order of the error $\psi^{(m)}$ and Lemma 2.9, we obtain the inequality:

$$|\xi_r^{(m)} - \xi_{r,\varepsilon}^{(m)}| \leq \sum_{k=0}^{m} \sum_{\mathbf{r}_i \in \Gamma_\varepsilon} C_m^k \Big(\prod_{j=0}^{i-1} Q_j \Big)^{(k)} (\tilde{C}_k h) \varepsilon^{1/2}.$$

The relations $h = C_h \varepsilon^{1/2}$ and $|\psi^{(k)}| \leq C_k \varepsilon$ allow us to choose the coefficient $\tilde{C}$, such that

$$\tilde{C}_k h \leq \frac{h}{\beta + \alpha\Delta + \alpha h} \tilde{C} e^{\lambda} + \tilde{\psi}^{(m-k)}(\mathbf{r}), \quad k = 0, ..., m.$$

Here $\tilde{\psi}(\mathbf{r})$ is the error function for problem (2.70) with $g \equiv 0$ and $\varphi \equiv \tilde{C} e^{\lambda}$. Hence, $|\mathrm{E}\xi_r^{(m)} - \mathrm{E}\xi_{r,\varepsilon}^{(m)}| \leq C_m \varepsilon^{1/2}$, where $C_m = \|u^{(m)}\| < \infty$, and u is the solution of problem (2.70) with $g \equiv 0$ and $\varphi \equiv \tilde{C} e^{\lambda}$. □

Theorem 2.18. *Under Lemma 2.10 conditions, the value* $\mathrm{E}(\xi_{r,\varepsilon}^{(m)})^2$ *is uniformly bounded:* $\mathrm{E}(\xi_{r,\varepsilon}^{(m)})^2 \leq D_m < +\infty$, $\varepsilon > 0$, $\mathbf{r} \in D$.

Proof. A set of the collision estimates $\{\xi_{r,\varepsilon}^{(i)}\}$, $i = 0, 1, ..., m$, according to system (1.59) can be represented in the following form (see Section 1.6):

$$\xi_0^{(i)} = \sum_{j=0}^{i} C_i^j q_1^{(i-j)} \xi_1^{(j)} + v_0, \quad i = 0, 1, \ldots, m, \tag{2.75}$$

where $\xi_t \equiv \xi_{r_t,\varepsilon}$, $v_0 \equiv v_{r_0}^{(i)}$ owing to the equality $g \equiv 0$. The weights satisfy the inequality

$$q_t^{(j)} = \frac{k^{(j)}(\mathbf{r}_{t-1}, \mathbf{r}_t, \lambda)}{p(\mathbf{r}_{t-1}, \mathbf{r}_t)} \leq 1, \tag{2.76}$$

if the transitions from $\mathbf{r}_{t-1}$ to $\mathbf{r}_t$ occur, and $q_t^{(j)} = 0$, if termination at this transition takes place.

In particular, for $m = 1$ system (2.57) takes the form:

$$\xi_0 = q_1 \xi_1 + v_0, \quad \xi_0^{(1)} = q_1 \xi_1^{(1)} + q_1^{(1)} \xi_1 + v_0^{(1)}. \tag{2.77}$$

Note, that the relations $\mathrm{E}(\xi_t^{(i)} \xi_t^{(j)}) < +\infty$ hold due to Theorem 1.14. Hence, the repeated averaging of the square of the first relation (2.77) results in the equation for $\mathrm{E}\xi^2$ (see Section 1.6):

$$\mathrm{E}\xi^2 = K_p(\mathrm{E}\xi^2) + v(2u - v).$$

Now, using (2.76) we obtain

$$\mathrm{E}\xi^2 \leq K(\mathrm{E}\xi^2) + C_v v,$$

i.e. $\mathrm{E}\xi^2 \leq C_u u$, and therefore the value $\mathrm{E}\xi^2$ is uniformly bounded.

Further, expression (2.12) and Lemma 2.9 ensure the inequality:

$$q_1^{(j)} \leq C_q v_{0,1}, \tag{2.78}$$

where $v_{0,1} = v_1(\mathbf{r}_0)$ corresponds to the equality $g \equiv 1$. Multiplying consecutively the second relation (2.77) by ξ_0 and $\xi_0^{(1)}$, after the repeated averaging of the obtained equalities with majorizing (2.78) we obtain the uniform boundedness of the values $\mathrm{E}(\xi\xi^{(1)})$ and $\mathrm{E}(\xi^{(1)})^2$. The further induction is obvious. □

3. The calculations of parametric derivatives for the problem

$$\Delta u - cu = 0, \quad \alpha u + \beta \frac{\partial u}{\partial \ell}\Big|_{\Gamma} = 1$$

realize the iterations of the operator $[\Delta - c]^{-1}$ under homogeneous boundary conditions. Indeed, for $u^{(1)}$ we have

$$\Delta u^{(1)} - cu^{(1)} = -u, \quad \alpha u^{(1)} + \frac{\partial u^{(1)}}{\partial \ell}\Big|_{\Gamma} = 0.$$

By induction, we then obtain

$$\Delta u^{(m)} - cu^{(m)} = -mu^{(m-1)}, \quad \alpha u^{(m)} + \beta \frac{\partial u^{(m)}}{\partial \ell}\Big|_{\Gamma} = 0, \ m = 1, 2, \dots.$$

Hence, in this case the following relation holds:

$$\frac{mu^{(m-1)}(\mathbf{r}_0)}{u^{(m)}(\mathbf{r}_0)} \underset{m\to\infty}{\longrightarrow} c^* + c, \quad \forall \mathbf{r}_0 \in D, \tag{2.79}$$

where $(-c^*)$ is the first eigenvalue of the Laplace operator for the domain D related to the boundary condition $\alpha u + \beta \frac{\partial u}{\partial \ell} = 0$.

Now let us construct algorithms for the calculation of parametric derivatives. The ε-biased estimator of the solution u is as follows (see Section 2.4):

$$\xi_\varepsilon = \sum_{i \in B_L} \Big[\prod_{j=0}^{i-1} Q(c, d_j)\Big] v(\mathbf{r}_i),$$

where $B_L = \{i : i = 1, \dots, L, \ \mathbf{r}_i \in \Gamma_\varepsilon\}$ is a set of the indices, and

$$Q(c, d_i) = \begin{cases} s(c, d_i)/s(c_0, d_i), & \mathbf{r}_i \in D \backslash \Gamma_\varepsilon, \\ 1, & \mathbf{r}_i \in \Gamma_\varepsilon. \end{cases}$$

Clearly, it is sufficient to construct algorithms to calculate the derivatives $Q_\lambda^{(m)}$ where

$$Q = \prod_{j=0}^{i-1} \frac{s(c, d_j)}{s(c_0, d_j)}.$$

Denote: $a = \sqrt{c}$. Differentiating initially with respect to the variable a:

$$Q'_a = Q(\ln Q)'_a,$$

it is easy to obtain the recurrent relation:

$$Q_a^{(m)} = \sum_{k=1}^{m} t_k Q_a^{(m-k)} \frac{(m-1)!}{(k-1)!(m-k)!}, \quad Q_a^{(0)} = Q,$$

where

$$t_k = (\ln Q)_a^{(k)} = \sum_{j=0}^{i-1} [\ln s(c, d_j)]_a^{(k)} = \sum_{j=0}^{i-1} t_k^{(j)},$$

and

$$Q^{(1)}_{|c|} = Q^{(1)}_a/(2a),\ Q^{(m)}_{|c|} = (2a)^{-m}\sum_{k=1}^{m} a^{(m)}_k a^{k-m} Q^{(k)},\ m > 1,$$
$$a^{(m)}_1 = (-1)^{(m-1)}(2m-3)!!,\ a^{(m)}_m = 1,$$
$$a^{(m)}_k = a^{(m-1)}_{k-1} - (2m-2-k)a^{(m-1)}_k,\ 1 < k < m.$$

The quantity $t^{(j)}_k$ is calculated for the jth sphere by using the following easy-to-derive formulae:

$$t^{(j)}_k = (k-1)!(-1)^{k-1}a^{-k} + d^k_j f_k,\ k = 1, 2, \ldots.$$

$$\begin{aligned} &f_1 = \mathrm{ch}(d_j a)/\mathrm{sh}(d_j a), && f_2 = f_1^2 - 1,\\ &f_3 = 2f_1 f_2, && f_4 = 2(f_2^2 + f_1 f_3),\\ &f_5 = 2(3f_2 f_3 + f_1 f_4), && f_6 = 2(3f_3^2 + 4f_2 f_4 + f_1 f_5),\\ &f_7 = 2(10 f_3 f_4 + 5 f_2 f_5 + f_1 f_6), && f_8 = 2(10 f_4^2 + 15 f_3 f_5 + 6 f_2 f_6 + f_1 f_7). \end{aligned}$$

At $c = 0$ we have

$$Q^{(m)}_c = \sum_{k=1}^{m} b_k t_k Q^{(m-k)}_c \frac{(m-1)!}{(k-1)!(m-k)!},\ Q^{(0)}_c = Q,$$

where

$$b_k t_k = \sum_{j=0}^{i-1} \lim_{c\downarrow 0}\,[\ln s(c, d_j)]^{(k)}_c = b_k \sum_{j=0}^{i-1} t^{(j)}_k,\ t^{(j)}_k = d^{2k}_j,$$

while the constant coefficients b_k can be obtained by an evident recursion from the representation

$$\ln\Big[\frac{d\sqrt{c}}{\sin(d\sqrt{c})}\Big] = -\ln\Big(\sum_{k=0}^{\infty}\frac{(-1)^k d^{2k}}{(2k+1)!}c^k\Big) = \sum_{p=1}^{\infty}\frac{(-1)^p}{p}\Big(\sum_{k=0}^{\infty}\frac{(-1)^k d^{2k}}{(2k+1)!}c^k\Big)^p = \sum_{k=1}^{\infty} d^{2k} b_k c^k/k!,$$

which holds for sufficiently small values of c. The first ten coefficients b_k are given below:

$$\begin{aligned} b_1 &= 1.666666667\cdot 10^{-1}, & b_6 &= 1.298642568\cdot 10^{-4},\\ b_2 &= 1.111111111\cdot 10^{-2}, & b_7 &= 7.893341227\cdot 10^{-5},\\ b_3 &= 2.116402116\cdot 10^{-3}, & b_8 &= 5.598081415\cdot 10^{-5},\\ b_4 &= 6.349206349\cdot 10^{-4}, & b_9 &= 4.537581858\cdot 10^{-5},\\ b_5 &= 2.565335899\cdot 10^{-4}, & b_{10} &= 4.137766635\cdot 10^{-5}. \end{aligned}$$

From the equality $s(0, \mathbf{r}) \equiv 1$, we have $\rho(K) < 1$ for the case $c = 0$ only if $\alpha(\mathbf{r}) > 0\forall\mathbf{r} \in \Gamma$. All the above statements hold when using $p(\mathbf{r}) = 1 - q(\mathbf{r}), \mathbf{r} \in \Gamma_\varepsilon$. Numerical results for the test Dirichlet problem are represented in Section 4.1 of [50].

4. Now, consider the problem

$$\Delta u + (\lambda + c)u = 0,\ \ \alpha u + \beta\frac{\partial u}{\partial \ell} = 1,\ \ u \in \Gamma, \tag{2.80}$$

where $c = c(\mathbf{r})$ and $0 \le -c(\mathbf{r}) - \lambda < c_0$. As well as in Item 3, differentiation of this problem with respect to λ results in iteration of the corresponding resolvent operator. Hence, the relation (2.79) holds with $-c^*$ equal to the first eigenvalue of the operators $\Delta u + c(x)$.

In [20], the following integral representation for equation (2.80) is constructed:

$$u(\mathbf{r}) = \frac{s(c_0,d)}{4\pi d^2}\int\limits_{S(\mathbf{r})} u(\mathbf{r}(\omega))d\omega + [1 - s(c_0,d)]\int\limits_{D(\mathbf{r})} p(\mathbf{r},\mathbf{r}')[c_0 + \lambda + c(\mathbf{r}')]c_0^{-1}u(\mathbf{r})d\mathbf{r}, \quad (2.81)$$

where $d = d(\mathbf{r}), \mathbf{r} \in D \setminus \Gamma_\varepsilon$, and

$$p(\mathbf{r},\mathbf{r}') = \frac{c_0 sh[(d - |\mathbf{r} - \mathbf{r}|)c_0]}{4\pi|\mathbf{r} - \mathbf{r}'|[sh(cd) - cd]},$$

Here $\int\limits_{D(\mathbf{r})} p(\mathbf{r},\mathbf{r}')d\mathbf{r}' = 1$. According to (2.81), in the proper walk on spheres and balls (with refection from the boundary) at the transition from a point $\mathbf{r}$, with the probability $s(c_0,d)$ a new point is uniformly chosen over the sphere $S(\mathbf{r})$ and with the probability $1 - s(c_0,d)$ within the ball $D(\mathbf{r})$, with respect to the density $p(\mathbf{r},\mathbf{r}')$, $\mathbf{r}' \in D(\mathbf{r})$. In the latter case the weight is multiplied by

$$[c_0 + \lambda + c(\mathbf{r}')]c_0^{-1} < 1. \quad (2.82)$$

It is known [20], that an average number of 'ball' states is less, than average number of 'sphere' states; therefore, the order of the computation cost here is the same as that in Section 2.4.

To correctly determine the collision estimate, it is necessary to introduce a new state coordinate, which possesses different values related to transitions to a ball or on a sphere. This coordinate determines a simulated density and a weight factor using the kernels from (2.81). From (2.82) for the new integral operator with $c_0 > 0$, the inequality $\rho(K) < 1$ also holds.

Note, that here it is possible to simulate termination with the probability $p(\mathbf{r}) \leq 1 - [c_0 + \lambda + c(\mathbf{r}')]c_0^{-1}$ at a 'ball point'. The value $\xi_r^{(m)}$ for the new algorithm is obviously computed and it satisfies Lemma 2.10 and Theorem 2.17. Theorem 2.18 here also holds because using (2.82) by analogy with [50], Section 3.4, we can show that $(\xi_r^{(m)})^2 \leq C_m\xi_r^{(m)}$. Note that this algorithm is applicable for $c \equiv \text{const}$ as well. But in this case the algorithm from Item 3 is more effective, especially for the Dirichlet boundary conditions, when it is applicable, also, for $c_0 = 0$.

Using relations (2.79) it is possible to estimate the moments $\mathrm{E}(\lambda^*)^\nu$, $\nu = 1, 2, \ldots$, when the functions $c(\mathbf{r}), \alpha(\mathbf{r}), \beta(\mathbf{r})$ and, perhaps, the boundary Γ, are random (see Section 3.4).

2.6. Global algorithms of the Monte Carlo method for solving n-dimensional difference equations

1. We analyze the Dirichlet problem for the Helmholtz differential equation,

$$\Delta u + cu = -g, \quad u|_\Gamma = \Psi, \quad (2.83)$$

in a domain $D \in R^n$ with a boundary $\Gamma; c(r) < c^*$, where $-c^*$ is the first eigenvalue of the Laplace operator defined on the domain D and $r = (x_1, \ldots, x_n) \in D$. We assume the regularity conditions are fulfilled for the functions c, g, and Ψ and for the boundary Γ, which guarantees the existence and uniqueness of a solution to the problem.

In the domain D, we construct a uniform grid with a mesh size h. As an estimate for the solution of the original problem at the grid points $r = (i_1 h, \ldots, i_n h)$, we consider the solution of the difference problem

$$\Delta_h u^h + c^h u^h = -g^h \quad \text{in} \quad D_h, \qquad a^h u^h = \Psi^h, \quad \text{on} \quad \Gamma_h, \tag{2.84}$$

where Δ_h is the conventional finite-difference analogue of the Laplace operator; D_h is the set of interior grid points; Γ_h is the set of grid points at which the boundary values are approximated; u^h is the grid function defined on $D_h \cup \Gamma_h, g^h$ and c^h are the values of the functions $g(r)$ and $c(r)$ at the grid points, respectively; a^h is a finite-difference approximation of the boundary conditions, such that the corresponding error is of the order of h^2; and Ψ^h is the value of the function $\Psi(\bar{r})$, where $\bar{r}$ is a certain point on Γ.

The properties of the finite-difference approximation of the Laplace operator suggest that all eigenvalues of the operator Δ_h defined on the domain D_h are negative when h is sufficiently small. We denote by $-c_h^*$ the eigenvalue that has the least absolute value. Then, problem (2.84) has a unique solution when $c^h < c_h^*$.

Equations (2.84) have the form

$$u^h_{i_1,i_2,..i_n} = \frac{u^h_{i_1-1,i_2,...i_n} + \cdots + u^h_{i_1,i_2,...i_n+1}}{2n\left[1 - (c^h_{i_1,i_2,...i_n} h^2)/(2n)\right]} + \frac{g^h_{i_1,i_2,...i_n} h^2}{2n - c^h_{i_1,i_2,...i_n} h^2}.$$

On the grid boundary, we set

$$u^h_{i_1,...,i_n} = \frac{\varepsilon u^h_{i_1,..i_k-1,..,i_n} + h\Psi(\bar{r})}{\varepsilon + h},$$

where $\bar{r}$ is the point of intersection of Γ with a grid line that is the nearest to a certain grid-boundary point, $u^h_{i_1,...,i_k-1,...,i_n}$ corresponds to the adjacent grid point lying on the same grid line, and ε is the distance from $\bar{r}$ to the grid-boundary point. This method of approximation of boundary conditions was proposed in [5].

Thus, we have the following set of linear algebraic equations:

$$u^h = Au^h + f^h, \tag{2.85}$$

We rewrite (2.85) in the form:

$$u_i^h = q_i \sum_{j=1}^{L} p_{ij} u_j^h + f_i^h, \tag{2.86}$$

where $i, j = (1, ..., L)$ are the numbers of grid points, $p_{ij} \geq 0$, and $\sum_{j=1}^{L} p_{ij} = 1$.

In accordance with the form of the set of equations, we have $p_{ij} = \frac{1}{2n}$, if i corresponds to an interior grid point and j corresponds to a grid point that lies next to it; $p_{ii} = 1$ at the grid-boundary points, and $p_{ij} = 0$ in other cases. At the grid-boundary points, we set $q_i = 0$; at the grid points that do not lie next to a boundary point, we set

$$q_i = \left[1 - c_i^h h^2/(2n)\right]^{-1};$$

at the grid points that lie next to boundary points, we substitute the boundary conditions to obtain

$$\bar{q}_i = q_i q_\varepsilon = \left(1 - \frac{c_i^h h^2}{2n}\right)^{-1} \left[1 - \frac{1}{2n}\frac{\varepsilon}{\varepsilon + h}\left(1 - \frac{c_i^h h^2}{2n}\right)^{-1}\right]^{-1}.$$

The absolute term f^h is defined as

$$f_i^h = \begin{cases} \dfrac{g_i^h h^2}{2n - c_i^h h^2}, & r_i \in D_h, \\ \dfrac{h}{\varepsilon + h}\Psi(\tilde{r}_i), & r_i \in \Gamma_h. \end{cases}$$

For simplicity, we henceforth consider the case when all grid-boundary points lie on the original boundary Γ; in other words, we assume that the domain D_h is the union of 'coordinate' parallelepipeds. In this case, we have

$$q_i = \left[1 - c_i^h h^2/(2n)\right]^{-1}, \tag{2.87}$$

$$f_i^h = \begin{cases} \dfrac{g_i^h h^2}{2n - c_i^h h^2}, & r_i \in D_h, \\ \Psi(\tilde{r}_i), & r_i \in \Gamma_h. \end{cases}$$

In what follows, we denote the spectral radius of matrix A by $\rho(A)$ and the matrix A with $c \equiv 0$ by A_0.

Lemma 2.11. *It holds that*

$$\rho(A_0) \leq 1 - c_h^* h^2/(2n).$$

Proof. The foregoing analysis shows that the set of equations

$$u^h = \left(1 - \frac{ch^2}{2n}\right)^{-1} A_0 u^h + f^h$$

with any f has a unique solution when c is constant and $-\infty < c < c_h^*$. It is well known that the spectral radius of a substochastic matrix A_0 corresponds to its positive eigenvalue λ_0. Assume that $\lambda_0 > 1 - c_h^* h^2/(2n)$. Then, there exists $\exists c < c_h^*$such that $1 - ch^2/(2n) = \lambda_0$ and the set of equations does not have solutions for some f, which leads to a contradiction.

By Lemma 2.11, for $c^h < c_{\max}^h < c_h^*$, we have

$$\rho(A) \leq \left[1 - \frac{c_{\max}^h h^2}{2n}\right]^{-1} \left[1 - c_h^* h^2/(2n)\right] < 1.$$

According to the theory of the Monte Carlo method [19], the corresponding solution to (2.86) can be represented as

$$u_{i_0}^h = \mathrm{E}\xi_{i_0} = \mathrm{E}\sum_{j=0}^{\tilde{N}} \left(\prod_{k=0}^{j-1} q_{i_k}\right) f_{i_j}^h,$$

where $i_0, \ldots, i_{\tilde{N}}$ are the numbers of the grid points that lie on the path of the Markov chain characterized by an initial distribution δ_{i_0} and transition probabilities p_{ij}, $\tilde{N}$ is the number of the step at which the walk first reaches the grid boundary. This estimate provides a probabilistic representation of the Neumann series for (2.85).

We note (see [2]) that estimates for solutions at various points can be based on a single path: a path is assumed to start from the ith state if this state is reached at a certain step of the random walk. Moreover, if the initial state is characterized by a uniform probability distribution over the boundary, the average number of arrivals at a points will be a constant number independent of the grid point (see Lemma 2.13 below).

Thus, a global estimate for the solution to (2.85) can be obtained by applying the following algorithm. The initial state of a Markov chain is chosen on the boundary with uniform probability. At the first step, the walk reaches the nearest interior grid point with unit probability; next, the path is continued to any adjacent grid point with equal probability. The chain is terminated when the path returns to the boundary. The estimate for the solution has the form

$$u_i = \mathrm{E}_i \sum_{j=m_i}^{\tilde{N}} q^{j-m_i} f_{i_j},$$

where the expected value is calculated by averaging over the paths containing the ith grid point, and m_i corresponds to the first arrival at that point (henceforth, we drop the index h and assume that $c \equiv \text{const} < c_h^*$ to simplify formulas without loss of generality).

Following [19], we refer to this method of solution as the *direct method*.

On the other hand, we can use the relation

$$u_i = (u, \delta_i) = (f, u^*),$$

where u^* is the solution to the conjugate equation

$$u^* = A^* u^* + \delta_i,$$

δ_i is the marker of a grid point:

$$\delta_i(j) = \begin{cases} 1, & \text{if} \quad i = j, \\ 0, & \text{if} \quad i \neq j. \end{cases}$$

By analogy with the algorithm described above, we have the following estimate for the solution:

$$u_i = \mathrm{E} \frac{f_{i_0}}{p_{i_0}} \sum_{j=0}^{\tilde{N}} q^j \delta_i(i_j),$$

where $i_0, \ldots, i_{\tilde{N}}$ are the numbers of the grid points that lie on the path of a walk characterized by a uniform initial distribution over the domain; i.e. $p_{i_0} \equiv p = 1/L$, where L is the number of grid points in the domain. This method for estimating the solution is known as the *conjugate method* (see [19]).

2. To analyze the variance of the direct method, we use the relation (see [19])

$$\mathrm{E}\xi_{i_0}^2 = \mathrm{E}\left(\sum_{j=0}^{\tilde{N}} q^j f_{i_j}\right)^2 = \mathrm{E}\left[\sum_{j=0}^{\tilde{N}} q^{2j} f_{i_j}(2u_{i_j} - f_{i_j})\right],$$

that is, $\mathrm{E}\xi_i^2 = (\chi, \delta_i)$, where χ is the Neumann series for the set of equations

$$\chi_i = f_i(2u_i - f_i) + \sum_{j=1}^{L} \frac{a_{ij}^2}{p_{ij}} \chi_j, \quad i = 1, \ldots, L.$$

First, we analyze the convergence conditions for the series. The inequality

$$q^2(c) = \left(1 - \frac{ch^2}{2n}\right)^{-2} \leq \left(1 - \frac{2ch^2}{2n}\right)^{-2} = q(2c)$$

implies that $E\xi_i^2$ is bounded from above by the solution of the set of difference equations (2.85) with f replaced by $|f|$ for $c' = 2c$, which uniformly converges to the solution of the corresponding boundary value problem as $h \to 0$. Therefore, when $c < c_h^*/2$, the variance of the direct method is uniformly bounded with respect to h:

$$D\xi_i \leq \text{const} + O(h^2).$$

To estimate the variance of the conjugate method, we use the relation (see [19])

$$E\xi_i^2 = E\left(\frac{f_{i_0}}{p_{i_0}} \sum_{j=0}^{L} q^j \delta_i(i_j)\right)^2 = E\left(\frac{f_{i_0}}{p_{i_0}}\right)^2 \sum_{j=0}^{L} q^{2j} \delta_i(i_j)[2u_{i_j}^* - \delta_i(i_j)] = (\chi, \delta_i(2u^* - \delta_i)),$$

where χ is the Neumann series for the set of equations

$$\chi_i = \frac{f_i^2}{p_i} + \sum_{j=1}^{L} \frac{a_{ij}^2}{p_{ij}} \chi_j, \quad i = 1, \ldots, L.$$

The series is convergent when $c < c_h^*/2$. Recalling that

$$f_i^2 = \begin{cases} \left(\dfrac{g_i h^2}{2n - ch^2}\right)^2 = O(h^2)\dfrac{h^2}{2n - ch^2}, & r_i \in D_h, \\ (\Psi_i)^2, \quad r_i \in \Gamma_h, \end{cases}$$

$$p_i \equiv p = 1/L = O(h^n),$$

we have

$$D\xi_i \leq \text{const}/h^{n-2} + O(h^2),$$

for $\Psi \equiv 0$ and $h \to 0$, whereas

$$D\xi_i \leq \text{const}/h^n + O(h^2),$$

for $\Psi \neq 0$.

Thus, the estimated variance of the conjugate method is $O(h^{-n})$.

3. A global estimate of the solution to problem (2.83) in the domain D can be obtained by estimating the solution of (2.84) at the grid points and constructing a complete $\tilde{u}(r)$ by linear interpolation. The corresponding computational cost is

$$S = NtE\tilde{N},$$

where N is the number of simulated paths required to reduce the error to a prescribed level, $E\tilde{N}$ is the mean path length, and t is the time required to simulate one transition (this quantity is an asymptotically linear increasing function of n in the limit $n \to \infty$). Note, that

$$\mathrm{E}\tilde{N} = \sum_{j=1}^{L} \mathrm{E}\nu_j,$$

where $\mathrm{E}\nu_j$ is the average of the number of arrivals at the jth point of a path.

Lemma 2.12. *If $p_0 \equiv p = 1/L$, i.e. the initial state of the chain that corresponds to a random walk on the grid is chosen with equal probability on $D_h \cup \Gamma_h$, then*

$$\mathrm{E}\tilde{N} = O(h^{-2}).$$

Proof. The average number of arrivals at the ith grid point can be represented as

$$\mathrm{E}\nu_i = \mathrm{E}\sum_{j=0}^{\tilde{N}} \delta_i(i_j) = \mathrm{E}\frac{p_0}{p_0}\sum_{j=0}^{\tilde{N}} \delta_i(i_j) = (u_2, \delta_i),$$

where u_2 is the solution of the finite-difference problem

$$u_2 = A_0 u_2 + \frac{1}{L}, \tag{2.88}$$

which is equivalent to problem (2.85) with $c \equiv 0$, and $f \equiv 1/L$. Recalling the assumption that D_h is the union of parallelepipeds, which was introduced to obtain (2.87), we have $L = \mathrm{const}h^{-n}$. Therefore, we can rewrite (2.88) as

$$\left(\frac{u_2}{h^{n-2}}\right) = A_0\left(\frac{u_2}{h^{n-2}}\right) + \mathrm{const}h^2.$$

As $h \to 0$, the solution of this set of equations uniformly converges to the solution of the boundary value problem for Poisson's equation $\Delta u = \mathrm{const}$, $u|_\Gamma = 0$. Consequently,

$$\mathrm{E}\nu_i = (u_2, \delta_i) = O(h^{n-2}),$$

$$\mathrm{E}\tilde{N} = \sum_{i=0}^{L} \mathrm{E}\nu_i = O(h^{-n})O(h^{n-2}) = O(h^{-2}).$$

□

Lemma 2.13. *Suppose the initial state of the Markov chain is chosen from the set of boundary points with equal probability: $p_i \equiv 0$ in D_h, and $p_i \equiv L_1^{-1} = C_0 h^{n-1}$ on Γ_h. Then, the average number of arrivals at any interior grid point is $C_0 h^{n-1}$, and $\mathrm{E}\tilde{N} = O(h^{-1})$.*

Proof. Under the assumptions of the lemma, the average number of arrivals at various interior grid points is given by the solution to the finite-difference problem $\Delta_h u_h = 0$, $u_h|_{\Gamma_h} = C_0 h^{n-1}$. Thus, the first assertation of the lemma is proved. Following the proof of Lemma 2.12, we obtain

$$\mathrm{E}\tilde{N} = \sum_{i=1}^{L} \mathrm{E}\nu_i = O(h^{-1}).$$

□

Now, let us estimate the errors of the methods. It is well known (see [5]) that, if $u(\mathrm{r}) \in C^4(D)$, then

$$|u(\mathbf{r}_i) - \mathrm{E}\xi_i| \le Ch^2;$$

therefore,

$$|u(\mathbf{r}) - \mathrm{E}\tilde{u}(\mathbf{r})| \le Ch^2,$$

where $\tilde{u}(\mathbf{r})$ is the complete solution obtained by linear interpolation of the estimate for the finite-difference solution based on N paths. This leads to the following respective $L_2(D)$-metric estimates for the errors of the direct and conjugate methods:

$$\mathrm{E}||u(\mathbf{r}) - \tilde{u}(\mathbf{r})||^2_{L_2} \le \int_D \mathrm{D}\tilde{u}(\mathbf{r})d\mathbf{r} + \int_D (u(\mathbf{r}) - \mathrm{E}\tilde{u}(\mathbf{r}))^2 d\mathbf{r} \le \frac{d_1}{N_1} + C^2h^4, \tag{2.89}$$

$$\mathrm{E}||u(\mathbf{r}) - \tilde{u}(\mathbf{r})||^2_{L_2} \le \frac{d_2}{Nh^n} + C^2h^4.$$

Here, the coefficient C is estimated in terms of the upper bound of the second-order derivatives of the solution. The coefficients d_i are the upper bounds of the variances of estimates; they are uniformly bounded with respect to h (see above) and are estimated by invoking the solutions of the corresponding boundary value problems with $c' = 2c$ and $f \equiv \text{const}$. By N_1, we denote the average number of paths that go through a grid point. It is obvious that $N_1 = N\mathrm{E}\nu_i^{(1)}$, where $\mathrm{E}\nu_i^{(1)}$ is the average number of first arrivals at the ith grid point, or, equivalently, the probability that a path goes through a certain grid point at least once.

Lemma 2.14. *It holds that*

$$\mathrm{E}\nu_i^{(1)} \ge (1 - F_n)\mathrm{E}\nu_i,$$

where F_n is the probability that a path returns to its starting point on the infinite grid.

Proof. It is obvious that

$$\mathrm{E}\nu_i = \mathrm{E}\nu_i^{(1)}/(1 - F_n)$$

on the infinite grid. Since every path is terminated on the boundary, the probability of return is less than F_n when the grid is bounded. □

It is clear that

$$\mathrm{E}\nu_i^{(1)} \asymp (1 - F_n)\mathrm{E}\nu_i,$$

as $h \to 0$ if the location of the ith grid point is fixed; this asymptotic equality holds uniformly for all grid points that do not belong to a certain neighbourhood of the boundary. Note that, by the well-known Pab theorem (see [22]), $F_n < 1$ for $n > 2$, and $F \approx 0.35$ for $n = 3$.

Lemma 2.14 guarantees that

$$N_1 = N\mathrm{E}\nu_i^{(1)} \ge (1 - F_n)N\mathrm{E}\nu_i \asymp Nh^{n-1}.$$

Substituting this expression into (2.89), we obtain the following estimate for the error of the direct method:

$$\mathrm{E}||u(\mathbf{r}) - \tilde{u}(\mathbf{r})||^2_{L_2} \le \frac{d_1}{Nh^{n-1}} + C^2h^4.$$

Errors of this kind cannot be estimated in terms of the metric of the space of continuous functions by invoking embedding theorems. Indeed, even if $u(\mathbf{r})$ is a sufficiently smooth function, the error of approximation of problem (2.83) by (2.84) is of the order of h^2. Therefore, knowing a solution of (2.84), we can estimate only the first derivatives of a solution of (2.83), whereas estimates for higher-order derivatives are required for embedding theorems to be applicable.

The minimization problem for the computational cost (the number of operations required for the estimated error of the direct method to be reduced to δ) has the form

$$S = Nt\mathrm{E}\tilde{N} \to \min_{N,h}, \quad \frac{d_1}{Nh^{n-1}} + C^2h^4 = \delta^2.$$

Solving this problem, we obtain the following results:

$$h^* = \left[\frac{n}{C^2(4+n)}\right]^{1/4} \delta^{1/2}, \quad N^* = \left[\frac{C^2(4+n)}{n}\right]^{(n-1)/4} \frac{d_1(4+n)}{4}\delta^{-(n+3)/2},$$

$$S^* = \left[\frac{C^2(4+n)}{n}\right]^{n/4} \frac{td_1(4+n)}{4}\delta^{-n/2-2}.$$

In the general case, the corresponding results obtained for the conjugate method are

$$h^* = \left[\frac{n+2}{C^2(6+n)}\right]^{1/4} \delta^{1/2},$$

$$N^* = \left[\frac{C^2(6+n)}{n+2}\right]^{n/4} \frac{d_2(6+n)}{4}\delta^{-n+4/2},$$

$$S^* = \left[\frac{C^2(6+n)}{n+2}\right]^{(n+2)/4} \frac{td_2(6+n)}{4}\delta^{-n/2-3};$$

for $\Psi \equiv 0$, we have

$$h^* = \left[\frac{n}{C^2(4+n)}\right]^{1/4} \delta^{1/2},$$

$$N^* = \left[\frac{C^2(4+n)}{n}\right]^{(n-2)/4} \frac{d_2(4+n)}{4}\delta^{-n+2/2},$$

$$S^* = \left[\frac{C^2(4+n)}{n}\right]^{n/4} \frac{td_2(4+n)}{4}\delta^{-n/2-2}.$$

Now, we proceed to the results of test calculations.

The direct and conjugate methods were applied to the Dirichlet problem in the cube $0 \le x_i \le 1 (i = 1, 2, ..., n)$

$$\Delta u + cu = 0, \quad u|_\Gamma = \cos(x_1\sqrt{c/n}) \cdots \cos(x_n\sqrt{c/n}).$$

Its exact solution has the form

$$u(\mathbf{r}) = \cos(x_1\sqrt{c/n})\cdots\cos(x_n\sqrt{c/n}).$$

We estimated the solution at the grid points lying on the line $x_2 = x_3 = ... = x_n$, $x_1 \in [0,1]$. The direct method was implemented in the following algorithm. The numbers of the grid points lying on a path were memorized, the path was walked backwards after reaching the boundary, and the corresponding weighted sum was calculated. In the conjugate method, the summation was performed in the course of the first passage. Thus, the computational cost per step in the walk is slightly higher for the direct method as compared to the conjugate method. However, this does not lead to any difference in the order-magnitude estimate of the computational cost obtained in the limit $\delta \to 0$. The Table 2.1 shows some results calculated for $h = 0.01$ and $N = 10^6$. Here, σ_1 and σ_2 are the estimated maximal rms deviations of the direct and conjugate methods, respectively; sig_1 and sig_2 are the corresponding values of maximal absolute error; T_1 and T_2 are the corresponding computing times measured in arbitrary units.

Our calculations have confirmed that the direct method is characterized by a smaller estimate variance and is more efficient than the conjugate method even when the required estimation accuracy is low.

Note that the weights employed should simply be differentiated m times with respect to c when $c = \text{const}$. It can easily be shown that the variances of estimates for derivatives of u_m^h thus obtained, for example, for the problem

$$\Delta_h u^h + cu^h = 0, \quad u^h|_\Gamma \equiv 1,$$

are finite. Therefore, we can readily implement the iterative process

$$mu_{m-1}^h/u_m^h \to c_h^* - c$$

to estimate the first eigenvalue of the difference Laplace operator. Moreover, it is well known that the functions u_m^h provide a solution to the Riquier problem for the difference metaharmonic equation.

Table 2.1

n	c	σ_1	sig_1	σ_2	sig_2	T_1	T_2
3	15	0.021	0.030	0.193	2.975	415	322
4	20	0.017	0.025	0.247	2.207	532	346
6	25	0.084	0.076	0.9102	1.130	635	395

It is quite clear that the algorithms considered and conclusion obtained here can be extended, by following Section 2.5, to systems of elliptic equations of the form

$$\Delta u_i(\mathbf{r}) + \sum_{j=1}^{m} c_{ij}(\mathbf{r})u_j(\mathbf{r}) = -g_i(\mathbf{r}), \quad u_i(\mathbf{r})|_\Gamma = \Psi_i(\mathbf{r}), \quad i = 1,2,...,m \tag{2.90}$$

without any substantial modification. The resulting efficiency can be particularly high in the case of a vector algorithm with matrix weighting; after each step, the weighting factor is multiplied by a matrix factor calculated, by analogy with q_i, in terms of the coefficients of the system. In this case, the vector estimate obtained by the direct method for the solution of system (2.90) has the form

$$u_{i_0} = \mathrm{E}\xi_{i_0} = \mathrm{E}\sum_{j=0}^{\bar{N}} \left(\prod_{k=0}^{j-1} Q_{i_k}\right) f_{i_j},$$

where $Q_s = [E - h^2(2n)^{-1}C_s]^{-1}$ is the matrix weighting factor, $C_s = \{c_{ij}(\mathbf{r}_s)\}$ is the matrix of coefficients of system (2.90) calculated at the sth grid point, and

$$f_s = \begin{cases} h^2(2nE - h^2C)^{-1}g_s, & \mathbf{r} \in D_h, \\ \Psi_s, & \mathbf{r} \in \Gamma_h. \end{cases}$$

A similar estimate is constructed by the conjugate method:

$$u_i = \mathrm{E}\xi_i = E\frac{f_{i_0}}{p_{i_0}}\sum_{j=0}^{\bar{N}} \left(\prod_{k=0}^{j-1} Q_{i_k}\right) \delta_i(i_j).$$

When h is sufficiently small, we have

$$\frac{h^2}{2n}\|C(\mathbf{r})\| < 1, \quad \forall \mathbf{r} \in \bar{D},$$

which entails the estimate

$$\|Q\| \le \left(1 - \frac{h^2}{2n}\|C\|\right)^{-1}.$$

Consequently, when the condition

$$\|C(\mathbf{r})\| < c_h^*, \quad \forall \mathbf{r} \in \bar{D},$$

is fulfilled, the results and estimates obtained above are extended to the vector case.

If the inverse matrix $[E - h^2(2n)^{-1}C]^{-1}$ is difficult to calculate, it can be randomized by using the conventional representation in the form of matrix progression, i.e. by replacing the matrix weight Q by its independent random unbiased estimate $\tilde{Q}$ (see [[19], Section 4.4]).

4. Here we shall use the process of a random walk on a grid with reflection in the boundary.

We construct a uniform grid with mesh-width h in the domain D and consider a solution to the following difference problem as an estimator for a solution of the original problem at the grid knots $r = (i_1h, ..., i_nh)$,

$$\begin{aligned} \Delta_h u^h - c^h u^h &= -g^h, \quad \text{in} \quad D_h, \\ a^h u^h &= \varphi^h, \quad \text{on} \quad \Gamma_h, \end{aligned} \tag{2.91}$$

where Δ_h is the standard difference analog of the Laplacian; D_h is the set of interior grid knots; Γ_h is the set of knots at which the boundary values are approximated; u^h is the grid function defined on $D_h \cup \Gamma_h$; g^h and c^h are the values of $g(\mathbf{r})$ and $c(\mathbf{r})$ at the grid knots; a^h is the difference approximation of the boundary conditions; and φ^h are the values of $\varphi(\bar{\mathbf{r}})$, where $\bar{\mathbf{r}}$ is a certain point of the boundary Γ.

For simplicity, henceforth we consider the case in which all boundary grid knots lie on the original boundary Γ; i.e. the domain D_h is the union of coordinate parallelepipeds. Moreover, we consider the boundary conditions on Γ of the form

$$\alpha u + \beta \frac{\partial u}{\partial \mathbf{n}} = \varphi,$$

where $\mathbf{n}$ is the outward normal to the boundary.

The system (2.91) in $D_h \setminus \Gamma_h$ has the form

$$u^h_{i_1,i_2,\dots,i_n} = \frac{u^h_{i_1-1,i_2,\dots,i_n} + \dots + u^h_{i_1,i_2,\dots,i_n+1}}{2n\left(1 + \frac{c^h_{i_1,i_2,\dots,i_n} h^2}{2n}\right)} + \frac{g^h_{i_1,i_2,\dots,i_n} h^2}{2n + c^h_{i_1,i_2,\dots,i_n} h^2}.$$

On the boundary of the grid domain Γ_h we put:

$$u^h_{i_1,\dots,i_n} = \frac{\beta^h_{i_1,\dots,i_n}}{\beta^h_{i_1,\dots,i_n} + \alpha^h_{i_1,\dots,i_n} h} u^h_{i_1,\dots,i_k-1,\dots,i_n} + \frac{h}{\beta^h_{i_1,\dots,i_n} + \alpha^h_{i_1,\dots,i_n} h} \varphi^h_{i_1,\dots,i_n}, \tag{2.92}$$

where $(i_1, \dots, i_k-1, \dots, i_n)$ is the adjacent knot of $i = (i_1, \dots, i_n)$ on the straight line whose direction coincides with the normal to the boundary at $\mathbf{r}_i$. This difference approximation of the boundary conditions provides an error of the first order; i.e. if $u(\mathbf{r}) \in C^4(D)$ then

$$|u(\mathbf{r}_i) - u^h(\mathbf{r}_i)| \le C_1 h. \tag{2.93}$$

Observe that, by analogy with Section 2.4, we can consider another way of approximating the boundary conditions, providing the error order h^2 in the estimator for a solution.

Thus, we arrive at the system of linear equations $u^h = Au^h + f^h$ or, in more detail,

$$u^h_i = q_i \sum_{j=1}^{L} p_{ij} u^h_j + f^h_i, \tag{2.94}$$

where $i, j = (1, \dots, L)$ is the number of grid knots, $p_{ij} \ge 0$, and $\sum_{j=1}^{L} p_{ij} = 1$. According to the form of the system, we have $p_{ij} = 1/2n$ if i is an interior knot and j is an adjacent knot. For the ith boundary knot, we have $p_{ij} = 1$ if j is an adjacent knot on the straight line along the normal at the boundary knot i and $p_{ij} = 0$ in the other cases. For the boundary knots we have

$$q_i = \frac{\beta^h_i}{\beta^h_i + \alpha^h_i h},$$

and for the others,

$$q_i = \left(1 + \frac{c^h_i h^2}{2n}\right)^{-1}.$$

The free term f^h is determined by the relation

$$f^h_i = \begin{cases} \dfrac{g^h_i h^2}{2n + c^h_i h^2}, & \mathbf{r}_i \in D_h, \\ \dfrac{h}{\beta^h_i + \alpha^h_i h} \varphi^h_i, & \mathbf{r}_i \in \Gamma_h. \end{cases}$$

It is clear that if the conditions

$$c(\mathbf{r}) \geq c_0 > 0, \quad \alpha(\mathbf{r}) \geq \alpha_0 > 0$$

are satisfied, then the following estimate holds for the spectral radius of the matrix A in (2.94):

$$\rho(A) \leq \max\Big(\Big(1 + \frac{c_0 h^2}{2n}\Big)^{-1}, \frac{\max \beta}{\max \beta + \alpha_0 h}\Big) < 1.$$

It is well known from the theory of the Monte Carlo methods that in this case a solution to (2.94) is representable as

$$u_{i_0}^h = \mathrm{E}\xi_{i_0} = \mathrm{E}\sum_{j=0}^{\tilde{N}}\Big(\prod_{k=0}^{j-1} Q_{i_k}\Big) f_{i_j}^h, \tag{2.95}$$

where $i_0, \ldots, i_{\tilde{N}}$ is the numbers of knots of the random trajectory of the Markov chain with the initial distribution δ_{i_0}, the transition probabilities p_{ij} and the termination probabilities $R_i \leq 1 - q_i$; $Q_i = q_i/(1 - R_i)$; and $\tilde{N}$ is the random number at which the trajectory terminates.

To estimate the solution at different points simultaneously, we may use the same trajectories, supposing that a trajectory starts at state i if it hits this state for the first time at some step of the random walk [2]. Therefore, we may propose the following algorithm. A trajectory starts in the domain $D_h \cup \Gamma_h$ in accord with the given initial distribution with probabilities

$$p_i \geq 0,\ i = 1, \ldots, L, \quad \sum_{i=0}^{L} p_i = 1.$$

We mark the grid knots through which the trajectory passes until the termination time. Passing backwards, calculate the estimator. The global estimator for a solution to (2.94) has the form

$$u_i^h = \mathrm{E}_i \sum_{j=m_i}^{\tilde{N}} \Big(\prod_{k=m_i}^{j-1} Q_{i_k}\Big) f_{i_j}^h,$$

where the mean value is calculated over the trajectories passing through the ith knot and m_i is the time of the first hitting of the knot.

To study the variance of the estimator, we use the following relation of [19]:

$$\mathrm{E}\xi_{i_0}^2 = \mathrm{E}\Big(\sum_{j=0}^{\tilde{N}}\big(\prod_{k=0}^{j-1} Q_{i_k}\big) f_{i_j}\Big)^2 = \mathrm{E}\Big(\sum_{j=0}^{\tilde{N}}\big(\prod_{k=0}^{j-1}(Q_{i_k})^2\big) f_{i_j}(2u_{i_j} - f_{i_j})\Big);$$

i.e. $\mathrm{E}\xi_i^2 = (\chi, \delta_i)$, where χ is the Neumann series for the system

$$\chi_i = f_i^h(2u_i^h - f_i^h) + \sum_{j=1}^{L} \frac{(q_i p_{ij})^2}{p_{ij}(1 - R_i)} \chi_j, \quad i = 1, \ldots, L.$$

Since $q_i/(1 - R_i) \leq 1$, we can conclude that $\mathrm{E}\xi_i^2$ is bounded from above by a solution of the system of difference equations of the from (2.94) which converges uniformly to a solution of the corresponding boundary value problem as $h \to 0$. Consequently, the variance of the estimator is uniformly bounded in h :

$$D\xi_i \le \text{const} + O(h).$$

For global estimation of a solution, we can estimate the values of a solution to the system at the grid knots and construct the linear interpolation of $\tilde{u}(\mathbf{r})$. The cost of the method is determined by the value $S = Nt\mathrm{E}\tilde{N}$, where N is the number of modeled trajectories needed for attaining the given error, $\mathrm{E}\tilde{N}$ is the mean length of a single trajectory, t is the time of modeling a single transition (an increasing function asymptotically linear in n as $n \to \infty$).

The mean length $\mathrm{E}\tilde{N}$ of a trajectory is determined by the choice of the initial distribution and the termination probabilities R_i. Consider two versions of this choice.

Lemma 2.15. *Suppose that $c(\mathbf{r}) \ge c_0 > 0$, and the Markov chain is determined by the following parameters: $p_i = 1/L$, i.e. the initial state of the 'walk on a grid with reflection' process is chosen uniformly in the domain $D_h \cup \Gamma_h$; $R_i = 1 - q_i$ in D_h, and $R_i = 0$ on Γ_h. Then $\mathrm{E}\tilde{N} = O(n/h^2)$ as $n \to \infty$ and $h \to 0$.*

Proof. The following formula for the mean length $\mathrm{E}\tilde{N}$ holds: $\mathrm{E}\tilde{N} = \mathrm{E}\tilde{N}_1 + \mathrm{E}\tilde{N}_2$, where $\mathrm{E}\tilde{N}_1$ is the mean number of visits to interior knots of D_h, and $\mathrm{E}\tilde{N}_2$ is the mean number of visits to boundary knots of Γ_h. Entering Γ_h, the next point is chosen in D_h. Hence, it is true that $\mathrm{E}\tilde{N}_2 \le \mathrm{E}\tilde{N}_1$. Let $i'_0, i'_1, \ldots$ are the numbers of knots of the random trajectory $i_0, i_1, \ldots$ excluding boundary knots. The following formula for $\mathrm{E}\tilde{N}_1$ holds:

$$\mathrm{E}\tilde{N}_1 = \sum_{j=0}^{\infty}\sum_{k=j}^{\infty}(1 - R_{i'_0}) \ldots (1 - R_{i'_{k-1}})R_{i'_k} = \sum_{j=0}^{\infty}(1 - R_{i'_0}) \ldots (1 - R_{i'_{j-1}}),$$

For the interior knots we have

$$q_i = \left(1 + \frac{c_i^h h^2}{2n}\right)^{-1} \le \frac{2n}{2n + c_0 h^2} = q.$$

Hence,

$$\mathrm{E}\tilde{N}_1 = 1 + \sum_{j=0}^{\infty} q_{i'_0} \cdots q_{i'_j} \le \sum_{k=0}^{\infty} q^k = \frac{1}{1-q}, \quad \mathrm{E}\tilde{N} \le \frac{2}{1-q} = \frac{2(2n + c_0 h^2)}{c_0 h^2}.$$

□

Lemma 2.16. *Suppose that $\alpha(\mathbf{r}) \ge \alpha_0 > 0$ and the initial state of the Markov chain is chosen equiprobably in the set of boundary points; i.e. $p_i = 0$ in D_h and $p_i = \dfrac{1}{L_1} = C_0 h^{n-1}$ on Γ_h, where L_1 is the number of boundary knots; $R_i = 1 - q_i$ for the boundary knots of Γ_h and $R_i = 0$ for the interior knots. Then $\mathrm{E}\tilde{N} = O(1/h^2)$ as $h \to 0$.*

Proof. The following estimate of a weight holds:

$$q_i = \frac{\beta_i^h}{\beta_i^h + \alpha_i^h h} \le \frac{\max\beta}{\max\beta + \alpha_0 h} = q, \quad \mathbf{r}_i \in \Gamma_h.$$

Further we will consider a version of the random walk on a grid with termination of a trajectory on Γ_h with probability $R = 1 - q$. The mean length of such random trajectory majorizes the mean length $\mathrm{E}\tilde{N}$ of the initial process trajectory because of the following formula for the mean number of visits to Γ_h :

$$\mathrm{E}\eta = 1 + \sum_{j=0}^{\infty}(1 - R_{i'_0}) \dots (1 - R_{i'_j}) \le \sum_{k=0}^{\infty} q^k = \frac{1}{1-q} = \mathrm{E}\eta_0,$$

where $i'_0, i'_1, \dots$ are the numbers of trajectory knots that belong to Γ_h, $\mathrm{E}\eta_0$ is the mean number of visits to Γ_h for the above modification of the process. The mean number of visits to the ith knot can be represented as

$$\mathrm{E}\nu_i = \mathrm{E}\sum_{j=0}^{\tilde{N}} \delta_i(i_j) = \mathrm{E}\frac{p_0}{p_0}\sum_{j=0}^{\tilde{N}} \delta_i(i_j) = (u_2^h, \delta_i),$$

where u_2^h is the solution of the difference problem

$$u_2^h = Au_2^h + p_0, \tag{2.96}$$

where the matrix A_0 is the matrix A with $c \equiv 0$, $\alpha \equiv \alpha_0$, $\beta \equiv \beta_{\max}$. The initial distribution is defined as

$$p_0(i) = \begin{cases} 0, & \mathbf{r}_i \in D_h, \\ 1/L_1, & \mathbf{r}_i \in \Gamma_h. \end{cases}$$

Taking into account the supposition about the 'parallelepipedness' property of D_h, we have $L_1 = C_3 \cdot h^{-n+1}$. The system (2.96) can be reduced to

$$\Big(\frac{u_2^h}{h^{n-2}}\Big) = A\Big(\frac{u_2^h}{h^{n-2}}\Big) + f^h,$$

where

$$f_i^h = \begin{cases} 0, & \mathbf{r}_i \in D_h, \\ C_2 h, & \mathbf{r}_i \in \Gamma_h. \end{cases}$$

The solution of this system uniformly converges to the constant solution of the boundary problem for the Poisson equation (as $h \to 0$):

$$\Delta u = 0, \quad \beta_{max}\frac{\partial u}{\partial n} + \alpha_0 u = C_4.$$

So we can conclude that $\mathrm{E}\nu_i = (u_2, \delta_i) = O(h^{n-2})$,

$$\mathrm{E}\tilde{N} = \sum_{j=0}^{\tilde{N}} \mathrm{E}\nu_j = O(h^{-n})O(h^{n-2}) = O(h^{-2}).$$

□

Now, let us estimate the mean error of the method. From (2.93) we obtain $|u(\mathbf{r}_i) - \mathrm{E}\xi_i| \le C_1 h$; consequently, $|u(\mathbf{r}) - \mathrm{E}\tilde{u}(\mathbf{r})| \le C_1 h$, where $\tilde{u}(\mathbf{r})$ is the linear interpolation of the estimator for a difference solution over N trajectories. Hence, we obtain the following estimate for the error in the metric of $L_2(D)$:

$$\mathrm{E}\|u(\mathbf{r}) - \tilde{u}(\mathbf{r})\|_{L_2}^2 \le \int_D \mathrm{D}\tilde{u}(\mathbf{r})d\mathbf{r} + \int_D (u(\mathbf{r}) - \mathrm{E}\tilde{u}(\mathbf{r}))^2 d\mathbf{r}.$$

For the method with absorption in the domain (Lemma 2.15), we have

$$\mathrm{E}\|u(\mathbf{r}) - \tilde{u}(\mathbf{r})\|^2_{L_2} \le \frac{d_1}{Nh^n} + C^2h^2,$$

and for the method with termination on the boundary (Lemma 2.16), the following inequality is satisfied:

$$E\|u(\mathbf{r}) - \tilde{u}(\mathbf{r})\|^2_{L_2} \le \frac{d_2}{N_1} + C^2h^2. \tag{2.97}$$

Here, the coefficients d_1 and d_2 depend on upper bounds of the variances of the estimators. The quantity N_1 is the mean number of trajectories passing through a single knot. Using the results of Item 3 for the Markov chain with initial uniform distribution over the boundary Γ_h and termination at the first hitting of the boundary, we have $N_1 \asymp Nh^{n-1}$. Inserting this expression in (2.96), we obtain the following estimate for the error for the method with termination on the boundary:

$$\mathrm{E}\|u(\mathbf{r}) - \tilde{u}(\mathbf{r})\|^2_{L_2} \le \frac{d_3}{Nh^{n-1}} + C^2h^2.$$

For the method with termination on the boundary, the problem of minimizing cost in the sense of the number of operations needed in order to attain the required error estimator δ has the form

$$S = Nt\mathrm{E}\tilde{N} \to \min_{N,h}, \quad \frac{d_3}{Nh^{n-1}} + C^2h^2 = \delta^2.$$

Solving this problem, we obtain the following values:

$$N^* = \frac{(n+3)d_3}{2}\Big(\frac{n+3}{n+1}C^2\Big)^{(n-1)/2}\delta^{-n-1}, \quad h^* = \frac{\delta}{C}\sqrt{\frac{n+1}{n+3}},$$
$$S^* = \frac{(n+3)d_3t}{2}\Big(\frac{n+3}{n+1}C^2\Big)^{(n+1)/2}\delta^{-n-3}.$$

For the method with termination in the domain, the corresponding values have the form

$$N^* = \frac{(n+4)d_1}{2}\Big(\frac{n+4}{n+2}C^2\Big)^{n/2}\delta^{-n-2}, \quad h^* = \frac{\delta}{C}\sqrt{\frac{n+2}{n+4}},$$
$$S^* = \frac{(n+4)d_1t}{2}\Big(\frac{n+4}{n+2}C^2\Big)^{(n+2)/2}\delta^{-n-4}.$$

Chapter 3

Parametric estimates for studying the radiation transfer in inhomogeneous media

3.1. Introductory information

A conventional mathematical model for a stationary radiation-transport process is the integro-differential transfer equation (see, for example, [14])

$$\begin{aligned}\omega\nabla\Phi + \sigma\Phi &= \int_V \sigma_s(\mathbf{r}, v')w_s(\mathbf{v}, \mathbf{v}'; \mathbf{r})\Phi(\mathbf{r}, \mathbf{v}')d\mathbf{v}' \\ &+ \int_V \nu(\mathbf{r}, v')\sigma_f(\mathbf{r}, v')w_f(\mathbf{v}, \mathbf{v}'; \mathbf{r})\Phi(\mathbf{r}, \mathbf{v}')d\mathbf{v}' + f_0(\mathbf{r}, \mathbf{v}) \qquad (3.1) \\ &\mathbf{r} \in D \subset R_3.\end{aligned}$$

Here, $\Phi = \Phi(\mathbf{r}, \mathbf{v})$ is the radiation intensity or the density of the flow of appropriate particles; V is the space of the velocities $\mathbf{v} = v \cdot \omega$; $v = |\mathbf{v}|$; $\sigma = \sigma(\mathbf{r}, v)$ is the total cross-section, $\sigma_s(\mathbf{r}, v)$ is the scattering cross-section, $\sigma_f(\mathbf{r}, v)$ is the fission (multiplication) cross-section, with

$$\sigma = \sigma_s + \sigma_f + \sigma_c,$$

where σ_c is the capture (disappearance of a particle) cross-section; $w_f(\mathbf{v}, \mathbf{v}'; \mathbf{r})$ is the indicatrix of fission; $\nu(\mathbf{r}, v')$ is the average number of secondary particles for each fission event initiated by a particle travelling at a speed v' at a point $\mathbf{r}$; and $f_0(\mathbf{r}, \mathbf{v})$ is the distribution density of the particle source.

With Monte Carlo methods, the natural boundary conditions for equation (3.1) are fulfilled automatically (see, for example, [41], [14]); however, if we assume that the medium is bounded by a convex surface and that outside the surface $\sigma = \sigma_c \neq 0$, then equation (3.1) can be extended to the entire space [41].

The radiation transport process can be considered to be a homogeneous Markov chain of collisions of a particle with elements of matter. A collision may result in the capture, scattering, or fission of the particle, with the probabilities being σ_c/σ, σ_s/σ, and σ_f/σ, respectively. The particle path terminates at the instant the particle is captured or escapes from the medium. If a multiplication takes place, we deal with a branching physical chain. The free path l of the particle between two successive collisions is distributed with the density

$$f_l(t) = \sigma(\mathbf{r}(t), v)\exp\left\{-\int_0^l \sigma(\mathbf{r}(t_1), v)dt_1\right\},$$

where $\mathbf{r}(t) = \mathbf{r}_0 + t\omega$, $\mathbf{r}_0$ is the initial point and ω is the path direction. This density is normalized provided that equation (3.1) is extended to the entire space in the above manner. The collision density $\varphi(\mathbf{r}, \mathbf{v})$ is related [14] to the density of the particle flux by

$$\varphi(\mathbf{r}, \mathbf{v}) = \sigma(\mathbf{r}, v)\Phi(\mathbf{r}, \mathbf{v}). \qquad (3.2)$$

The Monte Carlo method, i.e. the numerical simulation of the particle paths, can be used to estimate, for example, the following transport-process functionals. The integral

$$\int_{D_i}\int_V \varphi(\mathbf{r},\mathbf{v})d\mathbf{r}\,d\mathbf{v} \tag{3.3}$$

is the average number of collisions in a range D_i and estimated by the appropriate sample mean value. If the weight $1/\sigma(\mathbf{r},v)$ is assigned to the collisions, then by virtue of (3.2) we can estimate the integral at

$$J_i = J_{D_i} = \int_{D_i}\int_V \Phi(\mathbf{r},\mathbf{v})d\mathbf{r}\,d\mathbf{v}. \tag{3.4}$$

The estimate thus constructed is a special case of the so-called collision estimate.

It is also known (see, for example, [71]) that integral (3.4) equals the mean free path length L_i a particle travels in the range D_i, i.e. $J_i = \mathrm{E}(L_i)$. The corresponding Monte Carlo estimate is referred to as the free-path estimate (see Section 3.5).

Sometimes it is reasonable to transform equation (3.1) to an equivalent form by introducing fictitious 'delta-scattering'. Suppose that $\sigma(\mathbf{r},v) \le \sigma_m$. Add the equality

$$(\sigma_m - \sigma)\Phi = \int_V [\sigma_m - \sigma(\mathbf{r},v')]\delta(\mathbf{v}' - \mathbf{v})\Phi(\mathbf{r},\mathbf{v}')d\mathbf{v}'$$

to equation (3.1). The resultant equation can be treated as a transfer equation in a fictitious medium for which σ_m is the total cross-section, σ is the cross-section of physical collision, and $(\sigma_m - \sigma)$ is the cross-section of fictitious scattering with no change of velocity. In this case σ_m need not be constant. The straightforward simulation of the transport process thus described provides a correct simulation of the free path between physical collisions [44]. In some cases, such a transformation significantly simplifies the simulation of the free path in geometrically complex media. In the sequel, we will show that the use of delta-scattering can also simplify the calculation of parametric derivatives.

The collision density $\varphi(\mathbf{r},\mathbf{v})$ satisfies (see, for example, [41]) the integral equation

$$\varphi(x) = \int_X k(x'x)\varphi(x')dx' + f(x), \tag{3.5}$$

where $x = (\mathbf{r},\mathbf{v}) \in R_3 \times V$, f is the distribution density for primary collisions and $k(x',x)$ is the distribution density for the average number of secondary collisions that result from a single collision of a particle with a velocity $\mathbf{v}'$ at a point $\mathbf{r}'$, $\mathbf{x}' = (\mathbf{r}',\mathbf{v}')$. Similar to [41], it is easy to derive

$$\begin{aligned} k(x',x) &= \Big[\frac{\sigma_s(\mathbf{r}',v')}{\sigma(\mathbf{r}',v')}w_s(\mathbf{v},\mathbf{v}';\mathbf{r}') + \nu(\mathbf{r}',v')\frac{\sigma_f(\mathbf{r}',v')}{\sigma(\mathbf{r}',v')}w_f(\mathbf{v},\mathbf{v}';\mathbf{r}')\Big] \\ &\quad\times\frac{\sigma(\mathbf{r},v)\exp\{-\tau(\mathbf{r}',\mathbf{r};v)\}}{|\mathbf{r}-\mathbf{r}'|^2}\delta\Big(\omega - \frac{\mathbf{r}-\mathbf{r}'}{|\mathbf{r}-\mathbf{r}'|}\Big), \end{aligned}$$

where

$$\tau(\mathbf{r}',\mathbf{r};v) = \int_0^{|\mathbf{r}'-\mathbf{r}|}\sigma(\mathbf{r}'+t\omega)dt.$$

This expression is already connected with the weight modification of the process, namely, one secondary particle with the weight $\nu(\mathbf{r}',v')$is considered in the case of fission. In a more general weight algorithm (see, for example, [49]) the paths are simulated according to a certain transition density $p(x', x)$, with the auxiliary weight Q being multiplied by the weight factor $k(x', x)/p(x', x)$ after the transition $x' \to x$. However, in the multivelocity transport theory, the straightforward implementation of this algorithm is not possible due to incompatible singularities in the indicatrices of different types of scattering, because the particle energy upon scattering may be in a determinate way associated with the direction of scattering according to the scattering type. Therefore, in order to construct a general weight algorithm it is necessary to 'stratify' the general distribution of collisions according to their types by introducing the original number k of the collision type into the set of the coordinates in the phase space, i.e. put $x = (\mathbf{r}, \mathbf{v}, k)$, and $\mathrm{d}x = \mathrm{d}\mathbf{r}\mathrm{d}\mathbf{v}\mathrm{d}P(k)$. In this case the new value of x is defined after the collision type has been chosen according to the probabilities σ_k/σ, that is, when modelling a transition $x' \to x$ one should determine the new coordinate in the following succession: $\mathbf{v}$, $\mathbf{r}, k$.

The corresponding modified kernel of equation (3.5) takes the form:

$$k(x', x) = w_{k'}(\mathbf{v}, \mathbf{v}'; \mathbf{r}') \frac{\sigma(\mathbf{r}, v) \exp\{-\tau(\mathbf{r}', \mathbf{r}; v)\}}{|\mathbf{r} - \mathbf{r}'|^2} \frac{\sigma_k(\mathbf{r}, v)}{\sigma(\mathbf{r}, v)} \delta\Big(\omega - \frac{\mathbf{r} - \mathbf{r}'}{|\mathbf{r} - \mathbf{r}'|}\Big). \quad (3.6)$$

Here we may put $\sigma_1 \equiv \sigma_f$ and $w_1 \equiv \nu \cdot w_f$ and take into account various scattering types, including fictitious delta-scattering.

Note that this modification allows us to represent a physical particle source as the artifical collision source. Indeed, the origin of a particle trajectory can be regarded as the initial collision related to $k = 0$, resulting in scattering with a proper indicatrix (perhaps in δ-scattering).

The weight Monte Carlo algorithms are constructed by representing a solution to equation (3.5) by a Neumann series (see, for example, [41]). The availability of n_0 such that $\|K^{n_0}\| < 1$ will suffice for the series to converge. The operator norm is easily estimated in this case if the operator is considered to act from L_1 into L_1. The norm in L_1 is constructed with allowance made for the discreteness of the coordinate k.

In this case we have

$$\|K\| \le \sup\Big(\frac{\sigma_s}{\sigma} + \nu\frac{\sigma_f}{\sigma}\Big) = q.$$

Thus, if the branching is not involved in the simulation, in order for the Neumann series to converge it is logical to demand that the condition $q < 1$ be met.

The Monte Carlo method is commonly used to estimate linear functionals of the form:

$$J_h = (\varphi, h) = \int_X \varphi(x)h(x)\mathrm{d}x.$$

For instance, to estimate integral (3.3), put $h(x) = 1$, while to estimate integral (3.4), put $h(x) = 1/\sigma(\mathbf{r}, v)$ for $\mathbf{r} \in D_i$ and $h(x) = 0$ for $\mathbf{r} \notin D_i$. If $\{x_n\}$ is the physical chain of collisions (such a thing is possible within the framework of the model involved, provided $\nu \equiv 1$), then we have

$$J_h = \mathrm{E}\xi, \quad \xi = \sum_{n=0}^{N} h(x_n).$$

If a nonphysical chain of collisions is simulated with the transition density $p(x', x)$, then the collision estimate ξ is of the form:

$$\xi = \sum_{n=0}^{N} Q_n h(x_n),$$

where Q_n are auxiliary weights calculated by the relations:

$$Q_0 = \frac{f(x_0)}{\pi(x_0)}, \quad Q_n = Q_{n-1}\frac{k(x_{n-1}, x_n)}{p(x_{n-1}, x_n)}. \tag{3.7}$$

Here, $\pi(x)$ is the initial-state density being modelled. In this case the transition density $p(x', x)$ should include delta functions appearing in the appropriate indicatrices of scattering. The delta functions can be formally eliminated in the ratio $k(x', x)/p(x', x)$ and then the relation $\mathrm{E}\xi = J_h$ is justified in the regular way [41]. The weights are to be considered in more detail in Section 3.2 with the aim of constructing the estimates of parametric derivatives of the functionals like J_h.

For the variance of the random quantity ξ we have (see Section 1.4)

$$\mathrm{D}\xi = (\chi, h[2\varphi^* - h]) - (f, \varphi^*)^2,$$

where φ^* is the solution of the adjoint equation $\varphi^* = K^*\varphi^* + h$ and χ is the Neumann series for the equation

$$\chi(x) = \int_X \frac{k^2(x', x)}{p(x', x)}\chi(x')\mathrm{d}x' + \frac{f^2(x)}{\pi(x)}. \tag{3.8}$$

Thus, a simple modification of a physical branching transport process is the Markov chain for which the transition density $p_0(x', x)$ is derived from (3.6) by replacing $\nu \to 1$. Then Q_n is the product of the values of ν for the fission events preceding x_n, while the kernel of equation (3.8) is obtained from (3.6) by changing $\nu \to \nu^2$. Hence the variance of such a simple weight algorithm is finite if the following condition is met:

$$\sup\left(\frac{\sigma_s}{\sigma} + \nu^2\frac{\sigma_f}{\sigma}\right) = q_1 < 1. \tag{3.9}$$

It should be mentioned that an expression for the variance of the collision estimate, which is constructed on a branching Markov chain, is obtained in [7] (see also Section 1.6). It follows from the expression that in the case of direct simulation of a physical branching process the variance is finite and the value of J_h is also finite, i.e. for a subcritical system.

In conclusion, we note that restrictions imposed by condition (3.9) can be weakened if we increase the quantity σ_f by the term $\sigma_c = \sigma - \sigma_f - \sigma_s$ and recalculate the value of ν by the formula:

$$\nu' = \frac{\nu\sigma_f}{\sigma_f + \sigma_c} = \frac{\nu\sigma_f}{\sigma_a} \leq \nu.$$

In this case, kernel (3.6) and the function φ remain unchanged, while the paths in a finite system terminate with the probability of unity, due to the particle escape.

3.2. Calculation of parametric derivatives and critical values of parameters

1. In $L_\infty(X)$, we consider the following integral equation of the second kind with a kernel dependent on the parameter λ:

$$\varphi^*(x,\lambda) = \int_X k(x,y,\lambda)\varphi^*(y,\lambda)dy + h(x,\lambda) \tag{3.10}$$

which, in the operator notation, is of the form: $\varphi^* = K^*\varphi^* + h$. Here X is an n-dimensional Euclidean space. We pose the problem of the calculation of the derivatives:

$$\varphi^{*(n)}(x,\lambda) = \frac{\partial^n \varphi^*(x,\lambda)}{\partial \lambda^n}.$$

Let $K^{*(n)}$ denote an integral operator with the kernel $k^{(n)}(x,y,\lambda)$ and $\rho(K^*)$ stand for the spectral radius of the operator K^*.

Differentiating (so far only formally) relation (3.10) m times with respect to λ, we arrive at the triangular system of integral equations:

$$\varphi^{*(n)} = \sum_{i=0}^{n} C_n^i K^{*(n-i)}\varphi^{*(i)} + h^{(n)}, \quad n = 0,1,...,m \tag{3.11}$$

or $\mathbf{\Phi}^* = \mathbf{K}^*\mathbf{\Phi}^* + H$.

Vector Monte Carlo estimates have been considered in Section 1.9 for the solution of system (3.11). It has been shown that such estimates have finite variances if operators $K_p^{(n)}$ with the kernels $k^{(n)2}(x,y,\lambda)/p(x,y)$, $n = 1,...,m$, are bounded and the spectral radius of the operator K_p with the kernel $k^2(x,y,\lambda)/p(x,y)$ is less than unity. The above conditions agree with the standard conditions ensuring the finiteness of the variances of the scalar Monte Carlo estimates for equation (3.10). In Section 1.9 it also has been noted that the vector estimates of parametric derivatives are equivalent to the scalar estimates obtained by the straightforward differentiation of the collision estimates for the functionals involved. Thus, on the basis of the theory of vector estimates, a simple condition has been obtained for the finiteness of the variances of the scalar estimates of parametric derivatives which will be considered later.

Equation (3.10) is the conjugate of (3.5), but the results of this section are extended to the problems in Section 3.1, because the following representations are valid (see Section 1.4):

$$\xi = \frac{f(x_0)}{\pi(x_0)}\xi_{x_0}, \quad \mathrm{E}\xi_x = \varphi^*(x), \quad J_h = (f,\varphi^*) = (\varphi,h), \quad \mathrm{E}\xi^2 = \mathrm{E}\Big[\frac{f^2}{\pi^2}\mathrm{E}(\xi_x^2|x = x_0)\Big].$$

2. In this Item we will consider one of the cross-sections σ_k, σ in a certain subrange as a parameter λ.

As noted in Section 3.1, for the transfer equation involved, the relation $\rho(K) < 1$ holds provided $q < 1$ or if the medium is bounded, which results in a particle escaping from the medium with unit probability. Under these conditions, integral operators with the kernels of the form of $k^{(n)}(x,y,\lambda)$ are bounded and $\rho(K) < q_0 < 1$ if $\sigma_k \le C < +\infty$, $\sigma_0 = \sigma$. Moreover, for all σ_k' belonging respectively to the interval $\sigma_k \pm \varepsilon$, where ε is a sufficiently small negative number, the following inequality holds:

$$k(x,x') \le (1 + C_0\varepsilon)k_\varepsilon(x,x'),$$

where $k_\varepsilon(x,x')$ is the kernel of the transfer equation corresponding to $\sigma_k' - \varepsilon$, $k = 0,1,\ldots$. Thus, by virtue of Theorem 1.14 in order to construct the estimates of the derivatives

$\partial^m I_h/(\partial\sigma_k)^m$ we can perform direct differentiation of the estimate ξ. The variances of such estimates are finite if, for example, the simplest modification of direct simulation is performed (see Section 3.1) and condition (3.9) is met.

Now we will construct the scalar versions of the estimates of the derivatives of the functionals of the form of (3.3), (3.4) with respect to σ, σ_k, and σ_c.

First we estimate the derivative with respect to the total cross-section. If we assume σ_k to be constant, then we will actually estimate the derivative with respect to a scale factor ρ which is introduced for a certain subrange D_j by replacing σ with $\rho\sigma$ and σ_k with $\rho\sigma_k$. The paths are simulated for $\rho = \rho_0$. In this case the weight factor, according to (3.6), is of the form:

$$\frac{k(x',x)}{p_0(x',x)} = \nu_{k'}\left(\frac{\rho}{\rho_0}\right)^{\delta_j(r)} \exp\{-(\rho-\rho_0)\tau_j(\mathbf{r}',\mathbf{r};v)\}, \tag{3.12}$$

where $\nu_1 \equiv \nu$ and $\nu_k \equiv 1$ for $k \neq 1$; $\delta_j(r)$ is the indicator of the range D_j; $\tau_j(\mathbf{r}',\mathbf{r},v)$ is the optical path length in D_j from $\mathbf{r}'$ to $\mathbf{r}$ at $\rho = 1$. For definiteness, we will consider the derivative of functional (3.4), putting $h(x) \equiv \delta_i(r)$. From the above it follows that the estimate ξ can be differentiated here term by term, i.e. constructing the estimate for the quantity $J_j^{(m)}$ amounts to finding the expression for $Q^{(m)}$, where $Q \equiv Q_n$ is the current weight. On the basis of the representation

$$Q' = Q(\ln Q)'$$

we obtain

$$Q^{(m)} = \sum_{k=1}^{m} u_k Q^{(m-k)} \frac{(m-1)!}{(k-1)(m-k)!}\,. \tag{3.13}$$

It is obvious that

$$u_1 = (\ln Q)' = \frac{m_j}{\rho} - T_j, \quad u_k = (\ln Q)^{(k)} = \frac{(-1)^{k-1}}{\rho^k} m_j(k-1)!\,. \tag{3.14}$$

Here m_j is the total number of preceding collisions (including the current collision) in D_j, and T_j is a similar total optical path length in D_j for $\rho = 1$. Because we have to calculate sum (3.13), the highest derivatives can be estimated using a modification like the capture estimate (see, for example [49]). The estimate is calculated in this case only at capture, but with the additional weight σ/σ_c. Because the number of the collision type is considered to be an extra coordinate in the phase space, this estimate can be treated here as a special version of the collision estimate.

Now let us consider the estimate of the derivative with respect to the cross-section σ_k of some scattering type (or fission with fixed ν) and assume this quantity to be constant in D_j. Since the quantity $\sigma(\mathbf{r},v)$ is cancelled in (3.6), the quantities u_k are also determined by formulae (3.14) with the following substitutions:

$$\rho \to \sigma_k, \quad \rho_0 \to \sigma_k^{(0)}, \quad T_j = L_j,$$

where L_j is the total length of the previous path of a particle in D_j and m_j is the total number of the preceding scattering events of the k type in D_j.

Now let us estimate the derivative with respect to ρ when auxiliary fictitious delta-scattering is used (see Section 3.1). We assume that $\rho\sigma \leq \sigma_m$; the appropriate weight factor of the form:

$$\frac{k(x',x)}{p_0(x',x)} = \nu_k \cdot \left(\frac{\rho}{\rho_0}\right)^{\delta_j(r)}$$

for a physical collision and

$$\frac{k(x',x)}{p_0(x',x)} = \nu_{k'} \cdot \left[\frac{\sigma_m}{\sigma_m - \sigma\rho_0}\left(1 - \frac{\sigma}{\sigma_m}\rho\right)\right]^{\delta_j(r)}$$

for delta-scattering. The estimate by formulae (3.13) can be obtained here in the ordinary way; however, at σ/σ_m = const inD_j we can also calculate Q' immediately by differentiating the product

$$\rho^{m_j}\left(1 - \frac{\sigma}{\sigma_m}\rho\right)^{n_j}, \tag{3.15}$$

where m_j is the total number of physical collisions and n_j is the total number of delta-scattering events in the range D_j.

In calculating the derivatives with respect to σ_k, the weight factors are of the same form on substitution:

$$\sigma\rho \to \sigma_k, \quad \sigma\rho_0 \to \sigma_k^{(0)}$$

and if σ_k/σ_m = const in D_j, the quantity Q' can be calculated by differentiating the product

$$\sigma_k^{m_j}\left(1 - \frac{\sigma}{\sigma_m}\rho\right)^{n_j},$$

where m_j is now the total number of k-type scatterings in D_j. The expressions obtained should be used if auxiliary delta-scattering is employed to equalize cross-sections (see Section 1.3).

Let us mention a possible effective application of delta-scattering to stochastic problems. Assume that ρ is a random field and for simplicity assume that σ is a constant. One path of a particle is constructed for each realization ρ in estimating $\mathrm{E}\Phi_\rho$, the mean value of the solution when a double-randomization method is used. The weight is given by a more general expression than (3.15):

$$\left(\prod_{i=1}^{m_j} \rho_i\right) \prod_{i=1}^{n_j} \left(1 - \frac{\sigma}{\sigma_m}\rho\right)^{n_j}.$$

The appropriate collision estimate can be averaged (for a fixed path) over the field distribution ρ. Evidently, such averaging reduces to calculating power moments and requires much less information on the field ρ than its direct modelling, which, in addition, is not always possible. A similar procedure can be laid down for estimating other moments $\mathrm{E}\Phi_\rho^m$.

3. Let us assume now that a constant s is added to the total cross-section (and thereby to the capture cross-section $\sigma_c = \sigma - \sum_k \sigma_k$) in the range D_j. We find from (3.6) that

$$\frac{k(x',x)}{p_0(x',x)} = \nu_{k'} \cdot \exp\{-l_j s\},$$

where l_j is the path length in D_j for the corresponding free path. Let $\tilde{\nu}$ denote the product of the previous values of ν_k. Then we obtain

$$Q = \tilde{\nu} \exp\{-L_j s\},$$

where L_j is the total length of the previous path in D_j. Hence we have

$$Q^{(m)} = \tilde{\nu}(-L_j s)^m.$$

4. Let us write the integro-differential particle-transport equation in the form:

$$L\Phi + \sigma\Phi = S\Phi + \lambda S_f\Phi + f_0, \tag{3.16}$$

where (see Section 3.1) $L\Phi \equiv \omega\nabla\Phi$,

$$S\Phi \equiv \int_V \sigma_s(\mathbf{r}, v')w_s(\mathbf{v}, \mathbf{v}'; \mathbf{r})\Phi(\mathbf{r}, \mathbf{v}')d\mathbf{v}'$$

$$S_f\Phi \equiv \int_V \nu(\mathbf{r}, v')\sigma_f(\mathbf{r}, v')w_s(\mathbf{v}, \mathbf{v}'; \mathbf{r})\Phi(\mathbf{r}, \mathbf{v}')d\mathbf{v}'.$$

The principal (smallest in magnitude) eigenvalue $\lambda = \lambda_0$ is known to be positive under rather wide assumptions and can be found by the standard iteration method [14]. The quantity $1/\lambda_0 = k_{\text{eff}} = 1$ is called the effective multiplication factor of a system. If $k_{\text{eff}} < 1$, the system is subcritical; if $k_{\text{eff}} > 1$, it is supercritical; and at $k_{\text{eff}} = 1$ it is critical. Taking into account that Φ depends regularly on λ (see [14]), we differentiate (3.16) m times with respect to λ to obtain

$$L\Phi^{(m)} + \sigma\Phi^{(m)} = S\Phi^{(m)} + \lambda S_f\Phi^{(m)} + mS_f\Phi^{(m-1)} \tag{3.17}$$

or $R\Phi^{(m)} = mS_f\Phi^{(m-1)}$. Hence

$$\Phi^{(m)} = m!(R^{-1}S_f)^{(m-1)}\Phi'.$$

It has been shown that the operator $R^{-1}S_f$ has spectral properties [13] sufficient for the Kellog iterations to converge. Therefore we have

$$m\Phi^{(m-1)}/\Phi^{(m)} \to \lambda_1, \quad m \to \infty$$

and, in the limit, equation (3.17) takes the form:

$$L\Phi^* + \sigma\Phi^* = S\Phi^* + (\lambda + \lambda_1)S_f\Phi^*$$

i.e. $k_{\text{eff}} = (\lambda + \lambda_1)^{-1}$. Similarly, for linear functionals J of Φ of the form of (3.3), (3.4) we have

$$mJ^{(m-1)}/J^{(m)} \to \lambda_0 - \lambda = k_{\text{eff}}^{-1}(1 - \lambda k_{\text{eff}}).$$

Here the estimates of the quantities $J^{(m)}$ are made and justified by straightforward differentiation of the weight collision estimate (see Section 1.9)

$$\xi^{(m)} = \sum_{n=0}^{N}(\lambda^{m_n})^{(m)}Q_n h(x_n),$$

where m_n is the total number of the previous fission events. The estimate ξ is constructed on a modified path with $\lambda\nu \equiv 1$. The finiteness of the appropriate variances is ensured by the relation (see Section 3.1):

$$q_0 = \sup\left(\frac{\sigma_s}{\sigma} + \lambda^2\nu^2\frac{\sigma_f}{\sigma}\right) < 1.$$

Therefore it is reasonable to make equivalent modifications of the process by substituting

$$\sigma_f \to \sigma_f + \sigma_c, \quad \lambda\nu \to \frac{\lambda\nu\sigma_f}{\sigma_f + \sigma_c} < \lambda\nu.$$

Using the results of Section 1.6, we can show that the direct modelling of a branching physical process can be used to construct the estimates of $I^{(m)}$ with finite variances at $\lambda < k_{\text{eff}}^{-1}$. The algorithm of estimating k_{eff} has a number of advantages over the standard method of 'generations with a fixed number of points' [50], [19]. The estimates constructed with the weighted method or a correlated sampling can be realized by using the same ensemble of paths for different values of the system parameters. The estimate of the functional relations under study is thus effectively improved [26]. Using the standard procedure (Section 3.1 and [50]), we also can estimate the corresponding parametric derivatives and the perturbations of the quantities $I^{(m)}$ and, consequently, we can estimate k_{eff}. In the method of generations, such estimates are very hard to obtain [50].

5. Now let us consider the problem of estimating a time constant τ^* of particle multiplication, which determines the time exponential asymptotics $\exp(-\tau^* t)$ of the function Φ (see, for example, [14]). It is known [14] that if the quantity τ^*/v (v is the particle speed) is added to the absorption cross-section, the system becomes critical, i.e. τ^* is an eigenvalue of the equation that is derived from (3.16) by replacing

$$\sigma \to \sigma + \frac{\tau}{v}.$$

After the equation thus modified is differentiated m times, we obtain

$$L\Phi^{(m)} + \left(\sigma + \frac{\tau}{v}\right)\Phi^{(m)} = S\Phi^{(m)} + S_f\Phi^{(m)} - \frac{m}{v}\Phi^{(m-1)}$$

or $R_1\Phi^{(m)} = -(m/v)\Phi^{(m-1)}$. Hence

$$\Phi^{(m)} = (-1)^{(m-1)} m! (R_1^{-1} v)^{-(m-1)} \Phi'$$

and

$$m\Phi^{(m-1)}/\Phi^{(m)} \to \tau_1, \quad m \to \infty.$$

In the limit we have

$$L\Phi^* + \left(\sigma + \frac{\tau}{v}\right)\Phi^* = S\Phi^* + S_f\Phi^* - \frac{\tau_1}{v}\Phi^*$$

and $\tau^* = \tau + \tau_1$, i.e. we have here

$$mJ^{(m-1)}/J^{(m)} \to \tau^* - \tau.$$

It is easy to obtain the expression which is quite well known (see Item 3):

$$Q_n = \tilde{\nu}_n \exp(-\tau t_n),$$

where t_n is the total path time of a particle to r_n and $\tilde{\nu}_n$ is the weight, which takes into account fission events if the simplest modification of the process without branching is realized.

This implies that

$$Q_m^{(m)} = \bar{\nu}_n(-t_n)^m \exp(-\tau t_n),$$

i.e. if $\tau > 0$, the quantity τ_1 is negative and *vice versa*. Since the quantity τ/v is added to the capture cross-section, then the quantity q_0 modified by replacing $\sigma_f \to \sigma_f + \sigma_c$ can be made less than unity by choosing sufficiently large τ. It is very easy to justify here the unbiasedness of the estimates of the derivatives and the finiteness of their variances at $q_0 < 1$ on the basis of the vector approach. It has been mentioned above that the condition of the finiteness of the variances is less restrictive in modelling the branching. In addition to the above advantages over the method of generations, we now get another advantage: we need not interpolate by the values of $k_{\text{eff}}(\tau)$. The interpolation has serious drawbacks in this case [50], [19].

Now let us consider the estimate of the critical value of a scale factor ρ by which all the cross-sections in a certain range D_j are multiplied (see Item 2). On substituting

$$\sigma \to \rho\sigma, \quad \sigma_s \to \rho\sigma_s, \quad \sigma_f \to \rho\sigma_f$$

for $r \in D_j$ in equation (3.16) and on differentiating this equation m times with respect to ρ, we get

$$L\Phi^{(m)} + \rho T\Phi^{(m)} = -mT\Phi^{(m-1)},$$

where $T = \sigma - S - S_f$. Repeating the arguments of the preceding examples, we obtain the following limit equation:

$$L\Phi^* + (\rho + \rho_1)\sigma\Phi^* = (\rho + \rho_1)S\Phi^* + (\rho + \rho_1)S_f\Phi^*,$$

where $\rho_1 = \lim(m\Phi^{(m-1)}/\Phi^{(m)})$.

Hence we have here

$$mJ^{(m-1)}/J^{(m)} \to \rho^* - \rho.$$

Monte Carlo estimates for the quantities $I^{(m)}$ have been considered above. Although the quantity ρ has no effect on q_0, the variances of the collision estimates and hence the variances of the appropriate estimates of the quantities $I^{(m)}$ can be made finite by choosing a sufficiently small value of ρ, because a particle can disappear due to its escape from the system.

Similar to the above example, we can replace here the function Φ by a function $\varphi = \sigma\Phi$ because

$$\frac{m\varphi^{(m-1)}}{\varphi^{(m)}} = \frac{m[(m-1)\Phi^{(m-2)} + \rho\Phi^{(m-1)}]}{m\Phi^{(m-1)} + \rho\Phi^{(m)}} \underset{m\to\infty}{\to} \rho_1$$

and hence we can use functional (3.3) to construct computational estimates.

3.3. Use of the averaged estimates by the Monte Carlo method for the study of the effects of medium stochasticity

We consider the problem of estimating the probability of the passage and albedo of a particle for a parallelepiped medium, i.e. a finite two-dimensional layer of sufficiently large length. The density of the medium is a stochastic field. In order to solve this problem we use the weight estimates of the Monte Carlo method. We consider two models of a homogeneous stochastic medium, viz. a one-dimensional model and a three-dimensional one. We construct in the one-dimensional model the Poisson flux of points along the principal axis within the parallelepiped. As a result, the parallelepiped is divided into random layers. Then we independently choose in each layer the value of density which corresponds to the given one-dimensional distribution function. We construct in the three-dimensional model the Poisson fluxes along three principal axes within the parallelepiped. We independently choose in each random elementary parallelepiped the value of density according to the given one-dimensional distribution function. We perform the analytic partial averaging of the weight estimate with respect to the distribution of random density for fixed subdivision, and we obtain the finiteness conditions for the variance of such partially averaged estimates. We calculate the passage probabilities and the albedo probabilities. The weight algorithm allows us to calculate the small variations in the characteristics of the transfer process in passing from a deterministic medium to a stochastic one of the same average density, for example, the albedo variation. The weight algorithm also allows us to calculate the variation in the characteristics of the transfer process in passing from the one-dimensional model of a stochastic field to the three-dimensional one.

3.3.1. Modelling the homogeneous stochastic fields

We consider the models of homogeneous stochastic fields in the three-dimensional space related to point fluxes.

(1) We construct the Poisson point flux $\tau_0 = 0$, $\tau_{i+1} = \tau_i + \Delta\tau_i$, along the x-axis in the layer $\{0 \leq x \leq H, \quad (y,z) \in R^2\}$; here $\Delta\tau_i$ is a random quantity distributed with density $\lambda \exp(-\lambda t)$, $\lambda = L^{-1}$. The layer $\{0 \leq x \leq H, \quad (y,z) \in R^2\}$ is divided into m random layers $\tau_i \leq x \leq \tau_{i+1}$, $\tau_0 = 0$, $\tau_m = H$. Then we independently choose in each layer the value of σ_i, where

$$\sigma_i = \begin{cases} \sigma^{(1)} & \text{with probability } p, \\ \sigma^{(2)} & \text{with probability } 1-p. \end{cases}$$

The corresponding correlation function is equal to $\exp(-\lambda|\Delta x|)$ [50].

(2) We independently construct point fluxes of the above type along each axis for $0 \leq x \leq H_1$, $0 \leq y, z \leq H_2$. As a result, we get partitions:
along $x : (\tau_i, \tau_{i+1})$, $i = 0, 1, ..., m_x$, i.e. m_x layers;
along $y : (t_j, t_{j+1})$, $j = 0, 1, ..., m_y$, i.e. m_y layers;
along $z : (l_k, l_{k+1})$, $k = 0, 1, ..., m_z$, i.e. m_z layers.

Combining all the three partitions, we find that the parallelepiped $0 \leq x \leq H_1$, $0 \leq y, z \leq H_2$ is divided into $m_x m_y m_z$ subparallelepipeds. Further we independently chose in each elementary subparallelepiped the value

$$\sigma_{ijk} = \begin{cases} \sigma^{(1)} & \text{with probability } p, \\ \sigma^{(2)} & \text{with probability } 1-p. \end{cases}$$

Here the correlation function is equal to $\exp(-\lambda_x|\Delta x| - \lambda_y|\Delta y| - \lambda_z|\Delta z|)$ [50].

In what follows, these field models are used for solving the problems of the transport theory. In fact, we can consider the more general models of stochastic fields. We perform the above random partitions along the coordinate axes and independently choose in each layer (or a subparallelepiped) the random value of $\sigma_i(\sigma_{ijk})$ that corresponds to the given one-dimensional distribution function $F_\xi(x)$.

3.3.2. Partially averaged weight estimates

We designate: $r = (x, y, z)$, $X = (\mathbf{r}, \rho)$ is the collision point in the phase space of the coordinates and directions.

Let $\sigma(\mathbf{r})$ be the first one of the above stochastic fields, i.e. the one-dimensional Poisson field.

We consider the random quantity [41]

$$\xi = Q_N(\sigma) = \prod_{i=1}^{N} \frac{k(X_{i-1}, X_i, \sigma)}{p(X_{i-1}, X_i)} D_N, \quad Q_0 = 1,$$

where $k(X', X; \sigma)$ is the kernel of the integral equation of particle transport [41] with the parameters

$$\sigma_s(\mathbf{r}) = q\sigma(\mathbf{r}), \quad \sigma_c(\mathbf{r}) = (1-q)\sigma(\mathbf{r}), \quad 0 < q < 1, \tag{3.18}$$

and q is the survival probability of a particle at collision, i.e. the probability of its scattering according to a given indicatrix $w_s(\mu)$, μ is the cosine of the scattering angle. As the transitional density $p(X', X)$ of the modelled Markov chain of particle collisions we consider the kernel with the parameters

$$\sigma_{s,0} = q_0\sigma_0, \quad \sigma_{c,0} = (1-q_0)\sigma_0, \quad 0 < q_0 < 1, \tag{3.19}$$

and with the same indicatrix. The quantity D_N is zero or unity depending on the variant of escape whose probability is being estimated. Further, for the sake of simplicity the factor D_N will be omitted. The weight factor $k(X_{i-1}, X_i; \sigma)/p(X_{i-1}, X_i)$ for the given realization of $\sigma(\mathbf{r})$ is the density ratio of the free path probabilities upon scattering at the point X_{i-1} for media models with parameters (3.18), (3.19) and with the given indicatrix [41]. The sought probability

$$P = \mathrm{E}_\sigma \mathrm{M}_\omega \{Q_N(\sigma) | \sigma\}$$

is the mean value of the estimated functional for a stochastic medium (ω is a random trajectory)[41].

For the first model of the stochastic field in Section 3.3.1 we have

$$\begin{aligned} P = \mathrm{M}_{\omega,\{\tau_i\}} \Big\{ \prod_{i=1}^{m} \Big(p\Big(\frac{q\sigma^{(1)}}{q_0\sigma_0}\Big)^{n_i} \exp(-(\sigma^{(1)} - \sigma_0)l_i) \\ + (1-p)\Big(\frac{q\sigma^{(2)}}{q_0\sigma_0}\Big)^{n_i} \exp(-(\sigma^{(2)} - \sigma_0)l_i)\Big)\Big\}. \end{aligned} \tag{3.20}$$

Here, m is the number of random layers along x; n_i is the number of particle collisions in the ith layer; l_i is the free path of a particle in the ith layer; N is a random number of the final state of the collision chain, i.e. X_N is the absorption point or the first point of collision in the exterior of the domain that is assumed to be surrounded by a fictitious medium in which $\sigma = \sigma_0$ and $q = 0$.

For the second model of the stochastic field, formula (3.20) has the form

$$\mathrm{E}_\sigma \mathrm{M}_\omega\{Q_N(\sigma)|\sigma\}$$
$$= \mathrm{M}_{\omega,\{\tau_i\},\{t_j\},\{l_k\}}\Big\{\prod_{i=1}^{m_x}\prod_{j=1}^{m_y}\prod_{k=1}^{m_z}\Big(p\Big(\frac{q\sigma^{(1)}}{q_0\sigma_0}\Big)^{n_{ijk}}\exp(-(\sigma^{(1)}-\sigma_0)l_{ijk}) \tag{3.21}$$
$$+(1-p)\Big(\frac{q\sigma^{(2)}}{q_0\sigma_0}\Big)^{n_{ijk}}\exp(-(\sigma^{(2)}-\sigma_0)l_{ijk})\Big)\Big\},$$

where n_{ijk} is the number of particle collisions in the ijkth subparallelepiped; l_{ijk} is the path length in it; m_x is the number of layers along x; m_y is the number of layers along y; m_z is the number of layers along z.

As noted above, the distribution of the random quantity $\sigma_i(\sigma_{ijk})$ need not be the Bernoulli one. It may have any distribution function $F_\xi(x)$. If we can calculate analytically the expectation of $\mathrm{E}_\sigma\{Q_N(\sigma)|\omega,\{\tau_i\}\}$ for fixed trajectories and layer boundaries, we obtain the formulae analogous to (3.20) and (3.21). If we fail to analytically calculate the integral that is the expectation

$$\int_R \Big(\frac{q\sigma}{q_0\sigma_0}\Big)^{n_i}\exp(-(\sigma-\sigma_0)l_i)\mathrm{d}F_\xi(\sigma),$$

we can replace it by an approximate quadrature formula.

3.3.3. Finiteness conditions for the variance of a partially averaged weight estimate

The variance is finite if the quantity

$$\mathrm{E}_\sigma \mathrm{M}_\omega\{Q_N^2(\sigma)|\sigma\}$$

is finite, where

$$Q_N(\sigma) = \prod\Big(\frac{q\sigma}{q_0\sigma_0}\Big)^n\exp(-(\sigma-\sigma_0)l).$$

Here σ is a random value of density in the ijkth subparallelepiped, n is the number of particle collisions in the ijkth subparallelepiped, l is the path length in it.

$$Q_N^2(\sigma) = \prod\Big(\frac{q^2\sigma^2}{q_0^2\sigma_0^2}\Big)^n\exp(-2(\sigma-\sigma_0)l)$$
$$= \prod\Big(\frac{\overline{\sigma}}{q_0\sigma_0}\Big)^n\frac{\exp(-(\overline{\sigma}-\sigma_0)l)}{\exp(-(\overline{\sigma}-\sigma_0)l)}\exp(-2(\sigma-\sigma_0)l),$$

where we designate

$$\overline{\sigma} = \frac{q^2\sigma^2}{q_0\sigma_0}.$$

If the value of $\exp(-2(\sigma-\sigma_0)l)/\exp(-(\overline{\sigma}-\sigma_0)l)$ is no larger than unity, the mean value of $Q_N^2(\sigma)$ is no larger than that of the estimated functional for the section $\overline{\sigma}$ which is finite.

Thus, we must solve the inequality for σ:

$$\exp(-2(\sigma - \sigma_0)l + (\overline{\sigma} - \sigma_0)l) \leq 1,$$

that is equivalent to the inequality

$$2(\sigma_0 - \sigma) + (\overline{\sigma} - \sigma_0) \leq 0,$$

or, if we put $q = 0.9$, $q_0 = 1$, $\sigma_0 = q^2 = 0.81$ (see Section 3.3.5), to the inequality

$$\sigma^2 - 2\sigma + 0.81 \leq 0.$$

Hence, assuming that $q = 0.9$, $q_0 = 1$, $\sigma_0 = q^2 = 0.81$, we obtain the finiteness condition for the variance: $\sigma \in [0.6; 1.4]$.

3.3.4. Asymptotic estimation of the passage probability

Note that for a deterministic plane-parallel medium, the sought passage probability $I(H)$ may sometimes be estimated quite satisfactorily by the asymptotic formula

$$I_{as}(\tau(H)) \asymp e^{-\tau(H)/L}, \quad \tau(H) = \int_0^H \sigma(x)\mathrm{d}x, \tag{3.22}$$

where L is the diffusion length. If the scattering is highly anisotropic, in order to estimate the quantity L we can use the 'transport' approximation for the indicatrix

$$w(\rho, \rho') = (1 - \mu_0)\frac{1}{4\pi} + \mu_0\delta(\rho - \rho'). \tag{3.23}$$

This approximation retains the mean cosine of the scattering angle μ_0. The approximate value of L is determined by the relations

$$\frac{2\tilde{l}}{\tilde{q}L} = \ln\frac{L + \tilde{l}}{L - \tilde{l}}, \quad \tilde{q} = \frac{q(1 - \mu_0)}{1 - q\mu_0}, \quad \tilde{l} = \frac{l}{1 - q\mu_0}$$

For example, in the next section we obtain the estimate $I(20) \approx 0.0236$ by the Monte Carlo method for the radiation model with the standard Henyey-Greenstein indicatrix and parameters $\mu_0 = q = 0.9$, while formula (3.22) with 'transport' value $L = 5.4$ yields $I_{as}(20) = 0.0246$. It is clear that in this case we can improve the approximation (3.22), putting in front of the exponent the coefficient 0.0236/0.0246=0.959 for large values of $\tau(H)$.

Using the renewal theory and the asymptotic formula (3.22) the author obtained in [47], the asymptotic formula for the one-dimensional Poisson field $\sigma(x)$ with parameter λ (i.e. for the first variant in Section 3.3.1) becomes

$$\mathrm{E}I(\tau(H)) = \mathrm{E}I_{as}(\tau(H)) \asymp \frac{\lambda^{-2}}{\mathrm{E}(\lambda - \alpha + \sigma/L)^2}e^{-\alpha H}; \tag{3.24}$$

here the value of α is found from the equation

$$\lambda\mathrm{E}(\lambda - \alpha + \sigma/L)^{-1} = 1.$$

Indeed, using the designation

$$I(x,\sigma) = e^{-\tau(x)/L},$$

we first note that the conditional mean value of the quantity $I(x,\sigma)$ for the fixed value of x_1 in the Poisson model $\sigma(x)$ is

$$\mathrm{E}I(x,\sigma|x_1) = \begin{cases} \mathrm{E}e^{-\sigma_1 x_1/L}\mathrm{E}I(x-x_1,\sigma), & x_1 \le x \\ \mathrm{E}e^{-\sigma_1 x/L}, & x_1 > x \end{cases}$$

where σ_1 is a random value of $\sigma(x)$ in the interval (x_0, x_1). Averaging this expression over x_1, we get

$$\mathrm{E}I(x,\sigma) = \lambda \int_0^x \mathrm{E}e^{-\sigma_1 x_1/L} e^{-\lambda x_1}\mathrm{E}I(x-x_1,\sigma)\mathrm{d}x_1 + e^{-\lambda x}\mathrm{E}e^{-\sigma_1 x_1/L}. \tag{3.25}$$

This a special case of the so-called restoration equation. The asymptotic solution of (3.25) as $x \to \infty$ is given by the formula (see [22], p.443):

$$\mathrm{E}I(x,\sigma) \approx \frac{1}{\beta}e^{-\alpha x}\int_0^{+\infty} e^{\alpha t}e^{-\lambda t}\mathrm{E}e^{-\sigma_1 t/L}\mathrm{d}t = \frac{1}{\beta}\mathrm{E}(\lambda - \alpha + \sigma_1/L)^{-1}e^{-\alpha x}$$

where α is only root of the equation

$$\Psi(\alpha) = \lambda \int_0^{+\infty} e^{\alpha t}e^{-\lambda t}\mathrm{E}e^{-\sigma_1 t/L}\mathrm{d}t = 1 \tag{3.26}$$

and the quantity β is expressed by the formula

$$\beta = \lambda \int_0^{+\infty} te^{\alpha t}e^{-\lambda t}\mathrm{E}e^{-\sigma_1 t/L}\mathrm{d}t = \lambda\mathrm{E}(\lambda - \alpha + \sigma_1/L)^{-2}.$$

Equation (3.26) may be rewritten as

$$\lambda\mathrm{E}(\lambda - \alpha + \sigma_1/L)^{-1} = 1. \tag{3.27}$$

Thus, the asymptotic of the average intensity for the Poisson model $\sigma(x)$ is determined by the expression

$$\mathrm{E}I(x,\sigma) \approx \lambda^{-1}\frac{\mathrm{E}(\lambda - \alpha + \sigma_1/L)^{-1}}{\mathrm{E}(\lambda - \alpha + \sigma_1/L)^{-2}}e^{\alpha x} = I_0(x) \tag{3.28}$$

where α is determined from (3.27). Relations (3.27, 28) give us (3.24).

It is evident that $\Psi(\alpha)$ monotonically increases, and $\Psi(0) < 1$. Therefore the solution of (3.27) exists and is unique in $(0, \infty)$. Using the Jensen inequality (see, e.g. [22]) to estimate the quantity $\mathrm{E}e^{-\sigma_1 t/L}$ in (3.26), we obtain

$$\alpha < \mathrm{E}\sigma_1/L.$$

Using the results obtained, we can construct the asymptotic behaviour of the function $\mathrm{E}I(x,\sigma)$ for the generalized Poisson model $\sigma(x)$ as well, if we average (3.28) over λ with the weight $\mu(\mathrm{d}\lambda)$.

The numerical results in [47] show that the asymptotics (3.28) is generally a good approximation for $\mathrm{E}e^{-\tau(x)/L}$. We also note that this approximation is the one from below since $\mathrm{E}e^{-\tau(0)/L} = 1$, and from (3.27) we have

$$I_0(0) = \frac{\mathrm{E}(\lambda - \alpha + \sigma_1/L)^{-1}}{\lambda \mathrm{E}(\lambda - \alpha + \sigma_1/L)^2} = \frac{(\mathrm{E}(\lambda - \alpha + \sigma_1/L)^{-1})^2}{\mathrm{E}(\lambda - \alpha + \sigma_1/L)^2} < 1.$$

Averaging is performed here with respect to the one-dimensional distribution of the field σ, i.e. for the binary distribution we have $\mathrm{E}(\lambda - \alpha + \sigma/L)^{-1} = p(\lambda - \alpha + \sigma^{(1)}/L) + (1 - p)(\lambda - \alpha + \sigma^{(2)}/L)$.

We now consider possibilities of the analogous asymptotic estimates for the three-dimensional variant of the field $\sigma(x, y, z)$ in Section 3.3.1. It is clear that the large-scale horizontal [along (y, z)] inhomogeneity slightly affects the asymptotic behaviour, i.e. at $\lambda_y^{-1}, \lambda_z^{-1} >> L$ formula (3.24) with $\lambda = \lambda_x$ must give a satisfactory result in the three-dimensional case as well.

On the other hand, it is common knowledge that in the case of small-scale horizontal inhomogeneity (for $\lambda_y^{-1}, \lambda_z^{-1} << L$) the horizontal averaging is possible, i.e. the trajectory of a particle is actually constructed in the medium with $\sigma \equiv \mathrm{E}\sigma$, except for rigorously vertical directions.

Using the transport approximation (3.23), we can proceed to the consideration of the transport problem with isotropic scattering and the parameters

$$\sigma_{tr} = \nu\sigma, \quad \sigma_{s,tr} = q\sigma(1 - \mu_0), \quad q_{tr} = q(1 - \mu_0)/\nu, \quad \nu = 1 - q\mu_0.$$

After a particle falls vertically on the upper surface of a layer at the point ($x_0 = 0, y_0, z_0$) (the x-axis is assumed to be directed downwards), it can exprience several 'delta-scatterings' according to (3.23), after which it either escapes through the lower surface of the layer $x = H$ or is absorbed, or is scattered isotropically. The sought functional, i.e. the escape probability consists of two parts:

$$I^{(0)}(H) = I_1^{(0)}(H) + I_2^{(0)}(H),$$

where $I_1^{(0)}$ is the escape probability without isotropic scattering and absorption, and

$$I_2^{(0)}(H) = \int_0^H \mathrm{E}f(t;\sigma) i^{(0)}(t, H) \mathrm{d}t,$$

where $f(t;\sigma)$ is the distribution density of the first isotropic scattering for the given realization of σ, and $i^{(0)}(t, H)$ is the contribution to the sought functional, i.e. the escape probability for an isotropic single source of particles at the point $x = t,\ y = y_0,\ z = z_0$ for $\sigma \equiv \mathrm{E}\sigma$. Here the weight function $\mathrm{E}f(t, \sigma)$ may be approximately replaced by

$$f(t; \mathrm{E}\sigma) = \mathrm{E}\sigma_{s,tr} \exp(-t\mathrm{E}\sigma_{s,tr}),$$

i.e. by the distribution density of the first isotropic scattering in the modified medium with $\sigma \equiv \nu\mathrm{E}\sigma$. Thus, using the above effect of small-scale inhomogeneity on the particle transfer, we get the estimate

$$I_2^{(0)} \approx I_s^{(0)} - e^{\nu H \mathrm{E}\sigma}$$

where $I_s^{(0)}$ is the passage probability of a particle through the layer with $\sigma \equiv \mathrm{E}\sigma$ in the transport approximation. Using the asymptotics for $I_s^{(0)}$, we get

$$I_2^{(0)} \approx Ce^{-\nu H \mathrm{E}\sigma / L_0} - e^{\nu H \mathrm{E}\sigma},$$

where the coefficient C is close to unity and may be determined as described above. In the example with the Henyey-Greenstein indicatrix (see Section 3.3.5) we have $\nu = 0.19$, $q_{tr} = 0.09/0.19 \approx 0.4737$, $C = 0.959$ and $L_0 = 1.034$, i.e.

$$I_2^{(0)} = 0.959e^{-0.1838H} - e^{-0.19H}.$$

In order to determine the quantity $I_1^{(0)}(H)$ we note that when using the transport approximation we can express it by the formula

$$I_1^{(0)}(H) = \mathrm{E}e^{-\tau_{tr}^{(1)}(H)},$$

where

$$\tau_{tr}^{(1)}(H) = \int_0^H [\sigma_{s,tr}(t) + \sigma_{c,tr}(t)]\mathrm{d}t = \int_0^H \nu\sigma(t)\mathrm{d}t.$$

Thus, the quantity $I_1^{(0)}(H)$ may be asymptotically estimated by formula (3.24), replacing L by ν^{-1}.

As an example we consider the above problem (see also Section 3.3.5) for

$$\lambda^{-1} = L = 5.4, \quad H = 20, \quad p = 0.5, \quad \sigma^{(1)} = 0.6, \quad \sigma^{(2)} = 1.4, \quad \mathrm{E}\sigma = 1$$

Formula (3.24) in the transport approximation, i.e. for $L = 1.034/0.19$ has the form

$$\mathrm{E}I(\tau(H)) \asymp 0.8915e^{-0.1582H}.$$

The checking calculations by the Monte Carlo method show that it is advisable here to introduce an additional multiplier approximately equal to 1.035. The refined asymptotic formula is

$$\mathrm{E}I(\tau(H)) \asymp 0.9227e^{-0.1582H},$$

and for $H = 20$ we have $\mathrm{E}I(\tau(20)) \approx 0.03899$.

For the quantity $I_1^{(0)}(H)$ formula (3.24) yields the estimate

$$I_1^{(0)}(H) \asymp 0.8867e^{-0.1628H}.$$

Finally, we obtain for $I^{(0)}(H)$:

$$I^{(0)}(H) \approx 0.959e^{-0.1838H} - e^{-0.19H} + 0.8867e^{-0.1628H},$$

and $I^{(0)}(20) \approx 0.03610$, i.e. in this case the passage probability of a particle through a stochastic three-dimensional layer with small-scale horizontal inhomogeneity is 7 percent less than that for a stochastic horizontally homogeneous layer.

3.3.5. Test problem

We consider the problem of estimating the mean passage probability and the mean probability of the backward escape (albedo) of a particle for the layer of a substance: $0 \le x \le H_1$, $0 \le y, z \le H_2$, which is a stochastic field constructed for this layer in Section 3.3.1. We put

$$\lambda = 1/L, L = 5.4, H_1 = 20, H_2 = 40, p = 0.5, \sigma^{(1)} = 0.6, \sigma^{(2)} = 1.4.$$

Thus, the finiteness conditions for the variance of the weight estimate are satisfied (see Section 3.3.3).

We solve the problem by modelling the trajectories of particles by the standard scheme [41]. We model the value of the cosine of the scattering angle μ by the Henyey-Greenstein indicatrix [49]:

$$w_s(\mu) = \frac{1}{2}\frac{1-\mu_0^2}{(1+\mu_0^2-2\mu\mu_0)^{\frac{3}{2}}}, \quad -1 \le \mu \le 1, \quad \mu_0 = \mathrm{E}\mu = 0.9.$$

We use a weight algorithm. We model the trajectories of particle for the deterministic medium with constants $\sigma_0 = 0.81$ and $q_0 = 1$. The trajectories emerge from the point (0, 20, 20) along the x-axis and are observed up to the moment of leaving the domain: $0 \le x \le H_1$, $0 \le y, z \le H_2$. Outside the domain we have $q_0 = 0$.

We partition this domain into random parallelepipeds for each trajectory. We use the following algorithm for calculating the path lengths l_{ijk} in the subparallelepipeds.

Let l be a free path (chosen according to σ_0), (nx, ny, nz) is the number of a subparallelepiped in which a successive collision takes place, $(nx1, ny1, nz1)$ is the number of a subparallelepiped in which the preceding collision takes place. First we calculate the portions of the path length l in each layer:
along x : $lx[nx1], ..., lx[nx]$, the total sum is equal to l;
along y : $ly[ny1], ..., ly[ny]$, the total sum is equal to l;
along z : $lz[nz1], ..., lz[nz]$, the total sum is equal to l.

We further assign the current values of $nx1, ny1, nz1$ to the indices i, j, k, respectively. We compare three values: $lx[i]$, $ly[j]$, $lz[k]$. If $lx[i]$ is the least value, the current value of l_{ijk} increases by $lx[i]$. The current value of the index i also changes: if nx is larger than $nx1$, the value of i increases by unity, and if nx is less than $nx1$, the value of i decreases by unity. If $ly[j]$ or $lz[k]$ is the least value, the current value of l_{ijk} increases by $ly[j]$ or $lz[k]$, respectively. The current values of the indices j or k also change. If $nx1 = nx$ and $ny1 = ny$ and $nz1 = nz$, then l_{ijk} increases by l. The process ceases when $i = nx$, $j = ny$ and $k = nz$.

We have calculated the mean passage probability of a particle and the mean albedo probability for the above models of a stochastic medium.

We designate: N is the number of the trajectories modelled; $P^{(p)}$ and $P^{(a)}$ are the mean passage probability and the mean albedo probability for the media: $3s$ - a three-dimensional stochastic medium, $1s$ - a one-dimensional stochastic medium, $1d$ - a one-dimensional deterministic medium with attenuation coefficient $\sigma \equiv \mathrm{E}\sigma$; σ_N is the estimate of the mean square error.

Table 3.2 shows variation of the passage probability when passing from the one-dimensional stochastic medium to the three-dimensional one, and the corresponding mean square error. We see from Table 3.2 that the mean passage probability for the

Table 3.1. Calculation results of the mean passage probability.

N	$P^{(p)}_{3s}$	σ_N	$P^{(p)}_{1s}$	σ_N
10000	0,0325	0,0009	0,0350	0,0011
50000	0,0329	0,0004	0,0357	0,0005
100000	0,0338	0,0003	0,0366	0,0004
1000000	0,0340	0,0001	0,0368	0,0001

Table 3.2. Variation of the passage probability and the corresponding mean square error.

N	$P^{(p)}_{3s} - P^{(p)}_{1s}$	σ_N
10000	−0,0025	0,0005
50000	−0,0028	0,0003
100000	−0,0027	0,0002
1000000	−0,0028	0,00008

Table 3.3. Calculations results of the mean albedo probability.

N	$P^{(a)}_{3s}$	σ_N	$P^{(a)}_{1s}$	σ_N	$P^{(a)}_{1d}$	σ_N
10000	0,0696	0,0018	0,0699	0,0023	0,0700	0,0024
50000	0,0655	0,0008	0,0656	0,0010	0,0676	0,0010
100000	0,0647	0,0005	0,0653	0,0007	0,0671	0,0006
1000000	0,0659	0,0001	0,0667	0,0002	0,0676	0,0004

three-dimensional model of a stochastic field is about 8 percent less than that for the one-dimensional model.

We see from Tables 3.4 and 3.6 that there is a slight decrease in the mean albedo when passing from the deterministic medium to the stochastic one with the same average density: when passing to the three-dimensional model, by about 2.5 percent, and when passing to the one-dimensional model, by about 1.5 percent.

We see from Table 3.5 that when passing from the one-dimensional model of a stochastic medium to the three-dimensional one, the mean albedo decreases by about 1 percent.

An analysis of the estimates of the mean square errors for various values of N confirms the finiteness of the variances.

We also note that the estimate

$$P^{(p)}_{3s} - P^{(p)}_{1s} = -0.0028$$

with approximate mean square error $\sigma_N = 0.00008$ practically coincides with the analytic estimate

$$I^{(0)}(20) - \mathrm{E}I(\tau(20)) = -0.00289,$$

that is obtained in Section 3.4 for the case of small-scale horizontal inhomogeneity. This implies that the model considered in Section 3.3.4 is satisfactory even though $\lambda_y^{-1}, \lambda_z^{-1} \approx L$. The presence of inhomogeneity both along the y-axis and along the z-axis is evidently of importance here.

The albedo variation $P^{(a)}_{1d} - P^{(a)}_{1s} \approx 0.0009 > 0$ in Table 3.6 is quite logical since for the layers of large optical thickness τ the albedo P_a is obviously a convex function of the argument τ, i.e. $P_a'(\tau) > 0,\ P_a''(\tau) < 0$. Therefore its mean value is less than the function of the mean value of the argument.

Table 3.4. Albedo variations and the corresponding mean square error.

N	$P_{1d}^{(a)} - P_{3s}^{(a)}$	σ_N
50000	0,0020	0,0006
100000	0,0023	0,0003
1000000	0,0017	0,0001

Table 3.5. Albedo variations and the corresponding mean square error.

N	$P_{1s}^{(a)} - P_{3s}^{(a)}$	σ_N
50000	0,0001	0,0004
100000	0,0006	0,0003
1000000	0,0007	0,0001

Table 3.6. Albedo variations and the corresponding mean square error.

N	$P_{1d}^{(a)} - P_{1s}^{(a)}$	σ_N
50000	0,0020	0,0006
100000	0,0017	0,0004
1000000	0,0009	0,0001

The above models and the estimates are sufficiently versatile since calculations by the Monte Carlo method show that the mean value of radiation that passes through a stochastic optically thick layer is mainly determined by the correlation length of the medium that is equal to λ^{-1} for the Poisson field.

3.3.6. Additional remarks

(1) The calculation results in Section 3.3.6 confirm the high efficiency of the constructed aveaged estimates for calculating the variations of various radiation functionals when we change the probabilistic models of the medium. The computations performed show that the correlated sampling (e.g. with the use of the same random numbers [41], [49]) leads to lower accuracy under the conditions considered. However, this sampling is a direct modelling and always leads to estimates with finite variance when using double randomization for the direct unweighted estimate of the expectation $\mathrm{E}_\sigma M_\omega\{Q_N(\sigma)|\sigma\}$.

We can propose the readily realizable algorithm of the correlated sampling for the problems considered. We designate: $\boldsymbol{\rho} = (\rho_x, \rho_y, \rho_z)$ is a unit vector of the direction of the path, and

$$\tau(l) = \int_0^l \sigma(\mathbf{r} + \boldsymbol{\rho} t)\mathrm{d}t,$$

i.e. $\tau(l)$ is the optical path length from the point $\mathbf{r}$ in the direction $\boldsymbol{\rho}$. The direct modelling of the trajectory of a particle in an inhomogeneous medium consists in modelling successively the path length (by solving the equation $\tau(l) = -\ln\beta$, where β is a standard random number [41]), the survival or absorption at the end of the path, and in the case of survival, the new value of $\boldsymbol{\rho}$. It is obvious that when solving the radiation problems for various media with the same indicatrix and the survival probability q we can construct trajectories, using the premodelled universal random sequence

$$\{\ln\beta_i\},\ \{\boldsymbol{\rho}_i\},\quad i = 1,\dots,N,$$

that may be associated with an infinite homogeneous medium.

(2) Using the results obtained, we can state that for radiation calculations the stochastically inhomogeneous layer in the first approximation can be replaced by a homogeneous layer with complete attenuation coefficient $\sigma = \alpha L$, where L is the diffusion length for the given indicatrices of scattering and the survival coefficient q, and the value of α is determined from the one-dimensional Poisson approximation of the field according to (3.24). We should take into account that upon this replacement, the values of albedo at different levels are reconstructed adequately for the layer of sufficiently large optical thickness. If the horizontal inhomogeneity is essential, in the case of large optical thickness we can take into account only the first term in (3.29), i.e. we determine the value of α by the straight passing 'delta-scattered' radiation in the transport approximation.

We can refine the preaveraged constants, considering the results obtained by the Monte Carlo method for the original complicated stochastic models of media as experimental results. Then we can estimate the constants of the simplified models by the methods of solving the parametric inverse problems [41].

(3) Tables 3.2 and 3.4 show that introducing an additional horizontal inhomogeneity results in a decrease of the mean passage probability as well as of the mean albedo probability. The asymptotic estimates and additional computations for $H_2 = 100$ [56] gave us practically the same results, i.e. they related to the infinite (in horizontal directions) slab. The nature of the above double decreasing is explained by the effect of small-scale inhomogeneity to the particle transfer (see Section 3.3.4).

Only the very small value $P_{1d}^{(a)} - P_{1s}^{(a)}$ is not stabilized for $H_2 = 40$. It was especially estimated for the infinite slab. Using $N = 4000000$ here, the following estimate is obtained:

$$P_{1d}^{(a)} - P_{1s}^{(a)} \approx 0.0001.$$

3.4. Critical parameters of the particle transport process with multiplication in a stochastic medium

This Section deals with estimating the moments of the time constant τ and the efficient coefficient k of particle multiplication in a random scattering absorbing medium with particle multiplication. To do this, we derive the corresponding formulae for averaging the constants of the integro-differential transport equation, using a diffusion approximation as well as special iterative processes that are realized by the Monte Carlo method by calculating the corresponding parametric derivatives. We also consider the possibility of estimating the critical values of the parameters in the averaged solutions of the inhomogeneous particle-transport equation in a bounded stochastic medium, i.e. the values of the parameters τ and k for which the expectations of the solutions become infinite.

Moreover, we construct and evaluate the simple probabilistic models of the one-speed particle transport process, which is associated with the space averaging of the probability of particles escaping from a bounded medium. Besides being used for calculating the moments studied, these models turn out to be useful in the problem of determining the critical values of the parameters in the averaged solutions.

We give the computational results for homogeneous and two-layer balls with random values of density. These results show, in particular, the fast convergence of the mean values $\mathrm{E}\tau$ and $\mathrm{E}k$ to the corresponding critical values of the averaged transport equation as the correlation scale of a random medium decreases.

3.4.1. Averaging the constants and the solution of the transfer equation

For simplicity we consider the one-speed transport process with isotropic scattering and fission (multiplication) indicatrix. The values τ and k are determined by the relations [14], [40]:

$$(\omega, \mathrm{grad}\varphi) + \sigma\varphi - \sigma_s\hat{\varphi} - \nu\sigma_f\hat{\varphi} = -\tau\varphi \quad (T\varphi = -\tau\varphi), \tag{3.29}$$

$$(\omega, \mathrm{grad}\varphi) + \sigma\varphi - \sigma_s\hat{\varphi} = \frac{\nu\sigma_f\hat{\varphi}}{k} \quad (K\varphi = \frac{\nu\sigma_f\hat{\varphi}}{k}), \tag{3.30}$$

where

$$\hat{\varphi} = (4\pi)^{-1}\int \varphi(\mathbf{r}, \omega)\mathrm{d}\omega$$

is the total particle flux at the point $\mathbf{r}$, which is averaged over the directions ω. The eigenfunctions $\varphi(\mathbf{r}, \omega)$ are different in equalities (3.29), (3.30). The coefficients $\sigma(\mathbf{r})$, $\sigma_s(\mathbf{r})$, and $\sigma_f(\mathbf{r})$ are non-negative random fields, $\sigma_s + \sigma_f \leq \sigma$. The total number of particles at a fission point is $\nu \equiv \mathrm{const}$. The physical meaning of the values τ and k is that τ defines the time asymptotic behaviour of the number of particles in the system

$$\hat{n}(t) \asymp \exp(v\tau t)$$

where v is the particle speed and $k = k_{\mathrm{eff}}$ is the efficient multiplication coefficient in passing from 'generation' to 'generation'.

In particular, we consider the problem of calculating the mean values $\mathrm{E}\tau$ and $\mathrm{E}k$. In this section we average relations (3.29), (3.30), i.e. we determine the constant values $\sigma^{(0)}$, $\sigma_s^{(0)}$, and $\sigma_f^{(0)}$ of the coefficient in equalities (3.29), (3.30) for which $\mathrm{E}\tau$ and $\mathrm{E}k$ are eigenvalues (critical values). By analogy with [40] we use the corresponding averaged adjoint equations

$$T_0^*\varphi_0^* = -\varphi_0^*\mathrm{E}\tau, \quad K_0^*\varphi_0^* = \nu\sigma_f^{(0)}\hat{\varphi}_0^*/\mathrm{E}k. \tag{3.31}$$

If we multiply relation (3.29) by φ_0^*, the first equation in (3.31) by φ, and perform the element-by-element subtraction of the equalities obtained, we get the relation

$$(\varphi_0^*, [T - T_0]\varphi) = (-\tau + \mathrm{E}\tau)(\varphi_0^*, \varphi). \tag{3.32}$$

If we divide this relation by (φ_0^*, φ) and perform averaging, we get

$$(\varphi_0^*, \mathrm{E}\{[T - T_0]\varphi/(\varphi_0^*, \varphi)\}) = 0.$$

This equality holds if we put

$$\sigma^{(0)} = \mathrm{E}\left\{\frac{(\sigma\varphi_0^*, \varphi)}{(\varphi_0^*, \varphi)}\right\}, \quad \sigma_{s,f}^{(0)} = \frac{\mathrm{E}\{(\sigma_{s,f}\hat{\varphi}_0^*, \hat{\varphi})/(\varphi_0^*, \varphi)\}}{\mathrm{E}\{(\hat{\varphi}_0^*, \hat{\varphi})/(\varphi_0^*, \varphi)\}}. \tag{3.33}$$

Here we take into account that $(\hat{\varphi}_0^*, \varphi) = C(\varphi_0^*, \hat{\varphi})$.

Hence we can determine the constants of the operators T_0 and T_0^* from (3.33).

In order to determine the constant of the operators K_0 and K_0^* we multiply relation (3.30) by the corresponding function φ_0^* and the second relation in (3.31) by φ. The element-by-element subtraction of the equalities obtained yields the relation

$$(\varphi_0^*, [K - K_0]\varphi) = C\nu \left(\frac{(\sigma_f \hat{\varphi}_0^*, \hat{\varphi})}{k} - \frac{(\sigma_f^{(0)} \hat{\varphi}_0^*, \hat{\varphi})}{\mathrm{E}k} \right).$$

If we divide this relation by $(\hat{\varphi}_0^*, \hat{\varphi})/k$ and perform averaging, we get

$$(\varphi_0^*, \mathrm{E}\{k[K - K_0]\varphi/(\hat{\varphi}_0^*, \hat{\varphi})\}) = 0.$$

This equality holds if we set

$$\sigma_f^{(0)} = \mathrm{E}\{(\sigma_f \hat{\varphi}_0^*, \hat{\varphi})/(\hat{\varphi}_0^*, \hat{\varphi})\}.$$

Hence the necessary constant can be defined as

$$\sigma^{(0)} = \frac{\mathrm{E}\{k(\sigma\varphi_0^*, \varphi)/(\hat{\varphi}_0^*, \hat{\varphi})\}}{\mathrm{E}\{k(\varphi_0^*, \varphi)/(\hat{\varphi}_0^*, \hat{\varphi})\}}, \quad \sigma_s^{(0)} = \frac{\mathrm{E}\{k(\sigma_s \hat{\varphi}_0^*, \hat{\varphi})/(\hat{\varphi}_0^*, \hat{\varphi})\}}{\mathrm{E}k}$$
$$\sigma_f^{(0)} = \mathrm{E}\frac{(\sigma_f \hat{\varphi}_0^*, \hat{\varphi})}{(\hat{\varphi}_0^*, \hat{\varphi})}. \tag{3.34}$$

It is not difficult to see that for a determinate medium, formulae (3.33) and (3.34) coincide with each other and with standard homogenization formulae.

Note that functionals of the form

$$(\sigma_{s,f} \hat{\varphi}_0^*, \hat{\varphi}) = C^{-1}(\sigma_{s,f} \hat{\varphi}_0^*, \varphi)$$

can be calculated by using the standard 'collision estimate' of the Monte Carlo method (see Section 1.3). In order to estimate functionals of the form $(\sigma\varphi_0^*, \varphi)$ we can use the known relation [41]

$$\varphi_0^*(\mathbf{r}, \omega) = \mathrm{E}_{r,\omega} \hat{\varphi}^*(\mathbf{r} + \omega l),$$

where l is the random free path from the point $\mathbf{r}$ in the direction ω. It is advisable to randomize the corresponding collision estimate by replacing the values $\mathrm{E}_{r,\omega}\hat{\varphi}^*(\mathbf{r} + \omega l)$ by $\hat{\varphi}^*(\mathbf{r} + \omega l_{r,\omega})$, where $l_{r,\omega}$ are independently chosen auxiliary paths. In this case the mean value of the estimate does not change (see Section 1.7).

Let us consider the case in which the coefficients of equations (3.29), (3.30) are equal to the constants σ, σ_s, σ_f multiplied by the positive random quantity ρ that is uniformly distributed in the interval $(1 - \varepsilon, 1 + \varepsilon)$. From the above and simple analytic estimates we see that in this case the approximation can be efficient for a sufficiently solid medium:

$$(\varphi_0^*, \varphi) \approx C(1 - \gamma/\rho)(\hat{\varphi}_0^*, \hat{\varphi}), \tag{3.35}$$

where γ/ρ is the efficient averaged estimate of the probability of a particle escaping from the system (or the active section). We can estimate the coefficient γ by the preliminary direct calculations for a sufficiently small value of ε (see Section 2). Substituting (3.35) in (3.33) and (3.34), we obtain the expressions:

$$\sigma^{(0)} = \sigma \mathrm{E}\rho = \sigma, \quad \sigma_{s,f}^{(0)} = \sigma_{s,f} \frac{1+\gamma+(2\varepsilon)^{-1}\gamma^2 \ln[(1+\varepsilon-\gamma)/(1-\varepsilon-\gamma)]}{1+(2\varepsilon)^{-1}\gamma \ln[(1+\varepsilon-\gamma)/(1-\varepsilon-\gamma)]} \tag{3.36}$$

$$\sigma^{(0)} = \frac{\sigma[\mathrm{E}(k\rho) - \gamma \mathrm{E}k]}{\mathrm{E}k - \gamma \mathrm{E}(k/\rho)}, \quad \sigma_s^{(0)} = \sigma_s \frac{\mathrm{E}(\rho k)}{\mathrm{E}k}, \quad \sigma_f^{(0)} = \sigma_f. \tag{3.37}$$

The approximation $k \approx C\rho$ can be efficient here. If ρ is a random field, we can use the above procedure by replacing ρ by the random value

$$\rho_0 = C \int \rho(\mathbf{r}) \mathrm{d}\mathbf{r}$$

with corresponding distribution. Formulae (3.33), (3.34) show that for the small-scale ergodic field $\rho(\mathbf{r})$ the values $\mathrm{E}k$ and $\mathrm{E}\tau$ practically coincide with $k(\mathrm{E}\rho)$ and $\tau(\mathrm{E}\rho)$.

We briefly review the problem of determining the critical values of the parameters in the averaged solution of the inhomogeneous particle transport equation in the bounded medium:

$$(\omega, \mathrm{grad}\varphi_\rho) + (\rho\sigma + t)\varphi_\rho = \rho\sigma_s \hat{\varphi}_\rho + \nu\rho\sigma_f \hat{\varphi}_\rho / q + f_{tr},$$

i.e. such values of t (for $q = 1$) or q (for $t = 0$) that the expectation of the solution $\mathrm{E}\varphi_\rho(\cdot)$ becomes infinite.

This problem is most easily solved for a homogeneous medium with random density ρ which is distributed in the finite interval $(a, b]$ so that the corresponding density of probability distribution is continuous and does not vanish in the neighbourhood of the point $r = b$, i.e. in an interval $(b - \delta, b]$, $\delta > 0$. In this case the critical values of $\tau(\rho)$ and $k(\rho)$ are continuous rigorously monotonic functions of the density ρ, and for real geometries we have

$$\begin{aligned} \text{if } q = 1, \text{ then} \quad & \varphi_\rho(\cdot) < +\infty \text{ for } t > \tau(\rho) \\ \text{and} \quad & \varphi_\rho(\cdot) = +\infty \text{ for } t \le \tau(\rho); \\ \text{if } t = 0, \text{ then} \quad & \varphi_\rho(\cdot) < +\infty \text{ for } q > k(\rho) \\ \text{and} \quad & \varphi_\rho(\cdot) = +\infty \text{ for } q \le k(\rho). \end{aligned}$$

Proposition 1. *Under the above conditions the relations hold:*

$$\begin{aligned} \textit{if } q = 1, \textit{ then} \quad & \mathrm{E}\varphi_\rho(\cdot) < +\infty \textit{ for } t > \tau(b) \\ \textit{and} \quad & \mathrm{E}\varphi_\rho(\cdot) = +\infty \textit{ for } t \le \tau(b); \\ \textit{if } t = 0, \textit{ then} \quad & \mathrm{E}\varphi_\rho(\cdot) < +\infty \textit{ for } q > k(b) \\ \textit{and} \quad & \mathrm{E}\varphi_\rho(\cdot) = +\infty \textit{ for } q \le k(b). \end{aligned}$$

Proof. The inequality $\mathrm{E}\varphi_\rho(\cdot) < +\infty$ is obvious. On the other hand, if we have $t < \tau(b)$ for $q = 1$, because of the rigorous monotonicity of the function $\tau(\rho)$ we have

$$\exists\, \delta_1 > 0 \Rightarrow \varphi_\rho(\cdot) = +\infty, \quad \forall \rho \in (b - \delta_1, b],$$

With allowance for the continuity of $\mathrm{E}\varphi_\rho(\cdot)$ in t the first relation actually holds. The second relation is obtained analogously. □

The proved assertion is readily extended to the case of the piecewise constant random function $\rho(\mathbf{r})$ with independent values of the above type, i.e. the critical values of the parameters are equal to $\tau(b)$ and $k(b)$ for the case of the random medium as well.

It is quite clear that the inequality $E\varphi_\rho(\cdot) < +\infty$ with $q = 1$, $t > \tau(b)$ and $t = 0$, $q > k(b)$ also holds for the homogeneous random function $\rho(\mathbf{r})$. The converse relation can be obtained if there is a positive correlation between the values of ρ with nonzero correlation length $\hat{l}$. Thus, the discontinuity of the critical values of the parameters is quite possible, viz. the critical values are $\tau(b)$ and $k(b)$ $\forall \hat{l} > 0$, but these values coincide with $\tau(E\rho)$ and $k(E\rho)$ for $\hat{l} = 0$.

3.4.2. Use of the diffusion approximation

In the following, for a ball of radius R with the parameters

$$\rho\sigma, \quad \rho\sigma_s, \quad \rho\sigma_f, \quad \nu, \quad \sigma_s + \sigma_f = \sigma,$$

the values $E\tau$, $D\tau$ and Ek, Dk are calculated by the improved diffusion approximation [66] with extrapolated boundary:

$$\Delta\hat{\varphi} + \mu^2\hat{\varphi} = -\frac{g(r)}{D_0}, \qquad \hat{\varphi}(R + \alpha/p) = 0. \tag{3.38}$$

Here the random quantity $\rho > 0$ is uniformly distributed in the interval $(1 - \varepsilon, 1 + \varepsilon)$, $g(r)$ is the distribution density of an isotropic particle source, and $\alpha \approx 0.71044$. When calculating τ we take

$$p = \rho(\sigma_s + \nu\sigma_f), \quad D_0 = (p - \rho\sigma - \tau)/\mu^2$$

and when calculating k we take

$$p = \rho(\sigma_s + \nu\sigma_f/k(\rho)), \quad D_0 = (p - \rho\sigma)/\mu^2.$$

The value μ is obtained from the equations

$$\frac{p}{\mu}\operatorname{arctg}\left(\frac{\mu}{\rho\sigma}\right) = 1 \qquad \frac{p}{\mu}\operatorname{arctg}\left(\frac{\mu}{\rho\sigma + \tau(\rho)}\right) = 1 \tag{3.39}$$

when calculating k and τ, respectively. Since k and τ are the eigenvalues of equation (3.38) for $g(r) \equiv 0$, we have $R + \alpha/p = \pi/\mu$. Thus, we obtain from (3.39) the formula for the time constant:

$$\tau(\rho) = \frac{\pi p}{(Rp + \alpha)\tan(\pi/(Rp + \alpha))} - \sigma\rho,$$

and find the value $k(\rho)$ from the system

$$k(\rho) = \frac{\nu\rho\sigma_f}{p - \rho\sigma_s}, \quad \frac{Rp + \alpha}{\pi}\operatorname{arctg}\left(\frac{\pi p}{(Rp + \alpha)\rho\sigma}\right) = 1.$$

We carried out the main calculations for the values of the parameters:

$$\sigma = 1, \ \sigma_s = 0.97, \ \sigma_f = 0.03, \ \nu = 2.5, \ R = 7.72043.$$

The above formulae and the independent calculations show that in this case the value of R for $\rho = 1$ is quite close to the critical one, i.e. $k(1) \approx 1$ and $\tau(1) \approx 0$. The results obtained by numerical integration for different values of ε are presented in Tables 3.7, 3.8.

Table 3.7. Diffusion estimates for τ

ε	$\mathrm{E}\tau$	$\sqrt{\mathrm{D}\tau}$
0	0	0
0.1	−0.000110	0.004829
0.3	−0.001029	0.014796
0.5	−0.003098	0.025866

Table 3.8. Diffusion estimates for k

ε	$\mathrm{E}k$	$\sqrt{\mathrm{D}k}$
0	1	0
0.1	0.999062	0.061905
0.3	0.991831	0.184614
0.5	0.978792	0.303570

The results in Tables 3.7, 3.8 are checked by the Monte Carlo method (see Section 3.4.3). These results agree with the familiar fact that due to the stochasticity of the medium the average probability of a particle escaping increases (see, e.g. [3]). However, this dependence turns out to become weaker as the radius R of the ball decreases, viz. for $R = 5$ the value $\mathrm{E}k$ practically coincides with $k(\mathrm{E}\rho)$; for $R = 3$ we have $\mathrm{E}k > k(\mathrm{E}\rho)$, in particular, for $\varepsilon = 0.5$ we have $\mathrm{E}k - k(\mathrm{E}\rho) \approx 0.0061$, and in this case $k(\mathrm{E}\rho) \approx 0.3024$. It is interesting to note that $\mathrm{E}\tau < \tau(\mathrm{E}\rho)$ in all these versions.

Besides, we calculate the coefficient $\gamma = \gamma_1 \approx 0.0304$ in formula (3.35) such that the use of the corresponding constants (3.36) for $\varepsilon = \varepsilon_1 = 0.1$, $\gamma = \gamma_1$ ensures that the first equality in (3.31) holds for $\mathrm{E}\tau = \mathrm{E}(\tau|\varepsilon = 0.1)$ (see Section 3.4.5). Then for $\gamma = \gamma_1$ we calculate the constants (3.36) for $\varepsilon_2 = 0.3$ and $\varepsilon_3 = 0.5$ and the corresponding values of τ:

$$\tau_2 \approx -0.0011, \quad \tau_3 \approx -0.0034,$$

which are sufficiently close to the values in Table 3.7.

The analogous calculations with the use of the constants (3.37) and the approximation $k(\rho) \approx C\rho$ yield $\gamma = \gamma_1' \approx 0.0267$ and

$$k_2 \approx 0.9913, \qquad k_3 \approx 0.9741$$

which are close to the values in Table 3.8.

3.4.3. Estimation by the Monte Carlo method

We can obtain the estimates of the values $\mathrm{E}k$ and $\mathrm{E}\tau$ by the Monte Carlo method, using limiting relations of the form (see Section 3.2):

$$\lim_{m\to\infty} \frac{mJ_\tau^{(m-1)}}{J_\tau^{(m)}} = \tau - \tau_0, \qquad \lim_{m\to\infty} \frac{J_\lambda^{(m)}}{mJ_\lambda^{(m-1)} + J_\lambda^{(m)}/k_0} = k, \tag{3.40}$$

where J is the given linear functional of the particle flux φ; $J_\tau^{(m)}$, $J_\lambda^{(m)}$ are the m-order derivatives of this functional with respect to the parameters τ and $\lambda = 1/k$ in equations (3.29) and (3.30) with additional free members, i.e. in inhomogeneous problems with given particle sources; τ_0, k_0 are the values of the parameters for which the calculations are performed.

In the case of the homogeneous medium, in order to calculate the values of $J_\tau^{(m)}$ we model the nonphysical collision chain with the density of the initial state (a particle 'escapes' from the origin of coordinates)

$$\pi(x_0) = \frac{(\sigma+\tau_0)\exp\left(-(\sigma+\tau_0)\,r_0\right)}{r_0^2}\,\delta\!\left(\omega_0 - \frac{\mathbf{r}_0}{|\mathbf{r}_0|}\right)\frac{\sigma_{i_0}}{\sigma+\tau_0}$$

and with the transient (from x_1 to x) density

$$p(x_1, x) = \frac{\sigma_i}{\sigma + \tau_0} \frac{(\sigma + \tau_0) \exp(-(\sigma + \tau_0)|\mathbf{r} - \mathbf{r}_1|)}{4\pi |\mathbf{r} - \mathbf{r}_1|^2} \delta\left(\omega - \frac{\mathbf{r} - \mathbf{r}_1}{|\mathbf{r} - \mathbf{r}_1|}\right).$$

In this case the estimate of the functional $J = ((\sigma + \tau)\varphi, h)$ has the form (see Section 1.3):

$$\xi = \sum_{n=0}^{N} Q_n \, h(x_n).$$

Here $x = (\mathbf{r}, \omega, i)$, i is the number of the collision type (see Section 3.1), $r = |\mathbf{r}|$, Q_n are the auxiliary weights calculated by the formulae

$$Q_0(\tau) = \frac{f(x_0)}{\pi(x_0)}, \qquad Q_n(\tau) = Q_{n-1}(\tau) \frac{k(x_{n-1}, x_n)}{p(x_{n-1}, x_n)},$$

where $f(x)$ is the density of the first collisions, i.e.

$$f(x_0) = \frac{(\sigma + \tau) \exp(-(\sigma + \tau) r_0)}{r_0^2} \delta\left(\omega_0 - \frac{\mathbf{r}_0}{|\mathbf{r}_0|}\right) \frac{\sigma_{i_0}}{\sigma + \tau},$$

and $k(x_1, x)$ is the distribution density of the average number of secondary collisions after the single collision of a particle with direction ω_1 at the point $\mathbf{r}_1$ (see Section 3.1):

$$k(x_1, x) = \frac{\sigma_i}{\sigma + \tau} \tilde{\nu}_1 \frac{(\sigma + \tau) \exp(-(\sigma + \tau)|\mathbf{r} - \mathbf{r}_1|)}{4\pi |\mathbf{r} - \mathbf{r}_1|^2} \delta\left(\omega - \frac{\mathbf{r} - \mathbf{r}_1}{|\mathbf{r} - \mathbf{r}_1|}\right).$$

The value $\tilde{\nu}_1$ has the value of ν in the case of fission, otherwise it is equal to 1.

Thus, the 'collision point' x is defined by the space point $\mathbf{r}$, the direction ω of a colliding particle, and the number i of the collision type. We choose the new value of x by the scheme: $x_1 \to \omega \to \mathbf{r} \to i$, here multiplying the weight by the value $\tilde{\nu}_1$ is performed at the beginning of this procedure. It is easy to see that $\mathrm{E}\xi = ((\sigma + \tau)\varphi, h)$ does hold for $h \equiv h(\mathbf{r}, \omega)$, and if the density is $\rho \neq 1$, then $\mathrm{E}\xi = ((\rho\sigma + \tau)\varphi, h)$ since the collision density is $(\rho\sigma + \tau)\varphi$. The relationship of this type is acceptable in the iterative processes (3.40). We also note that the considered density $f(x_0)$ corresponds to the particle source that is concentrated at the origin and hence it is independent of the parameters as assumed when justifying the convergence in (3.40)(see Section 3.2).

Using simple transformations, we can obtain the representation for the weights:

$$Q_0(\tau) = \exp(-(\tau - \tau_0) r_0), \qquad Q_n(\tau) = \tilde{\nu}_n \exp(-(\tau - \tau_0) t_n),$$

where $\tilde{\nu}_n = \nu^{j_n}$, j_n is the number of fissions before the nth collision, t_n is the maximum range of particle before the nth collision.

We can obtain the estimates of the derivatives by differentiating the weight statistical estimates of the functional J. The conditions for the nonbias and finiteness of the variance of the estimates thus obtained coincide with the corresponding conditions for the initial estimate of the type ξ (see Section 3.2). By m-fold differentiation we get

$$Q_0^{(m)}(\tau_0) = r_0^m (-1)^m, \qquad Q_n^{(m)}(\tau_0) = \tilde{\nu}_n (-1)^m t_n^m.$$

We can decrease the variance of the weight estimate if we model the process without absorption but with the replacement

$$\sigma_f \to \sigma_f + \sigma_c, \qquad \nu \to \frac{\nu\sigma_f}{\sigma_f + \sigma_c}.$$

When calculating the values $J_\tau^{(m)}$ we use this transformation for $\sigma_c = \tau_0$.

Analogously, when calculating the values $J_\lambda^{(m)}$ for $\tau = \tau_0 = 0$ we obtain the expressions

$$Q_0^{(m)}(\lambda_0) = 1, \quad Q_n^{(m)}(\lambda_0) = \begin{cases} \tilde{\nu}_n j_n! \lambda^{j_n - m}/(j_n - m)!, & m \le j_n, \\ 0, & \text{otherwise.} \end{cases}$$

In order to estimate the values $J_\tau^{(m)}$, $J_\lambda^{(m)}$ we use expressions of the form

$$\xi^{(m)} = \sum_{n=0}^{N} Q_n^{(m)} h(x_n).$$

It is not difficult to see that the values $Q_n^{(m)}$ are determined by the above expressions in the case of the inhomogeneous medium, i.e. for $\rho = \rho(\mathbf{r})$. When calculating the values $J_\lambda^{(m)}$ we model branching with a coefficient ν_0 in order to decrease the variance of the estimates.

Note that in order to average the prelimiting expressions of the form (3.40) over ρ, in this case it is impossible to directly employ the efficient 'double randomization' method (see Section 1.7). In the framework of this method we model only a small number of conditionally independent trajectories of particles for the chosen realization of the random coefficients σ, σ_s, σ_f. In order to overcome this difficulty we can approximate the fraction x/y from (3.40):

$$\frac{x}{y} \approx \frac{x_0}{y_0} + \sum_{n=1}^{M} (-1)^{n+1} \frac{(y - y_0)^{n-1}}{y_0^{n-1}} \left(\frac{x}{y_0} - \frac{x_0}{y_0} \cdot \frac{y}{y_0} \right). \tag{3.41}$$

Here, y_0 is the defined function of the realizations of σ, σ_s, σ_f This function can be obtained by filling the set of the pre-estimated values of $\{y_0(\rho^{(i)})\}$ in a suitable way, where ρ is the defined functional of σ, σ_s, σ_f (it can be a vector functional), which largely determines the critical value of the parameter. Thus, the problem considered reduces to the estimation of moments of the form $\mathrm{E}(xy^n)$. We can obtain this estimate by the 'double randomization' method by modelling a group of $(n+1)$ conditionally independent trajectories for each realization of σ, σ_s, σ_f since the relation

$$\mathrm{E}(\zeta \prod_{k=1}^{n} \eta_k | \sigma, \sigma_s, \sigma_f) = xy^n$$

holds, where ζ, $\{\eta_k\}$ are the conditionally independent statistical estimates of the values x and y, respectively. Analogously we estimate the values $\mathrm{E}\tau^2$ and $\mathrm{E}k^2$.

We have obtained the estimates of the values $\mathrm{E}\tau$ and $\mathrm{E}k$ for the problem in Section 3.4.2 and for the analogous problem with piecewise constant random density ρ by the Monte Carlo method. We choose the values of the density ρ, which are uniformly distributed in the interval $(1-\varepsilon, 1+\varepsilon)$, independently for each of two parts of equal volume, which are obtained when the ball of radius $R = 7.72043$ is divided by the sphere of radius $R/2^{1/3}$. The functional $J = ((\rho\sigma + \tau)\varphi, h)$ is defined by the function $h(r) = \sin(\chi r)/r$. This function is a diffusion approximation to the eigenfunction of the homogenized problem for

$$\chi = \chi(\rho_0), \qquad \rho_0 = \int \rho(\mathbf{r})\,\mathrm{d}\mathbf{r}/(4\pi R^3)$$

and

$$\chi(\rho_0) = \frac{\pi\rho_0(\sigma_s + \nu\sigma_f)}{R\rho_0(\sigma_s + \nu\sigma_f) + \alpha} \quad \text{or} \quad \alpha \operatorname{arctg}\left(\frac{\chi(\rho_0)}{\sigma\rho_0}\right) = \pi - R\chi(\rho_0)$$

when calculating τ or k, respectively. The isotropic point particle source is located at the centre of the ball. We put $x_0 \equiv x_0(\rho), y_0 \equiv y_0(\rho)$ in the approximation (3.41) for the homogeneous ball, the values of $x_0(\rho)$ and $y_0(\rho)$ being calculated by quadratic interpolation. The values

$$J(\tau_0), \quad J(k_0), \quad \tau_m(\rho) = \frac{J_\tau^{(m-1)}}{J_\tau^{(m)}}, \quad k_m(\rho) = \frac{J_\lambda^{(m)}}{(mJ_\lambda^{(m-1)} + J_\lambda^{(m)}/k_0)}$$

are pre-estimated at three points $1-\varepsilon$, 1, $1+\varepsilon$. We can obtain the values of $J(\tau_0)$, $J(k_0)$, $\tau_m(\rho)$ and $k_m(\rho)$ at any point ρ by quadratic interpolation. Therefore we find successively the derivatives of the functional for any ρ by the formulae

$$J_\tau^{(m)} = \tau_m(\rho)J_\tau^{(m-1)}, \quad J_\lambda^{(m)} = \frac{mk_m(\rho)k_0J_\lambda^{(m-1)}}{k_0 - k_m(\rho)}, \quad m = 1,2,3,4. \tag{3.42}$$

We use the vector functional $\rho = (\rho_1, \rho_2)$, for the inhomogeneous ball, where ρ_1 is the random density in the ball $r \le R\cdot 2^{-1/3}$ and ρ_2 is the random density in the layer $R\cdot 2^{-1/3} \le r \le R$. The values of $J(\tau_0)$, $J(k_0)$, $\tau_m(\rho)$, and $k_m(\rho)$ are pre-estimated at five points, viz. at the vertices of the square $(1\pm\varepsilon, 1\pm\varepsilon)$ and at its centre $(1,1)$. We calculate these values for the arbitrary value of ρ by bivariate linear interpolation on the basis of the corresponding triangulation. We calculate the derivatives of the functional by formulae (3.42).

Table 3.9. Estimates for the homogeneous ball ($m = 2$)

ε	$\bar{\tau}_N \pm \sqrt{\mathrm{D}\bar{\tau}/N}$	$\bar{k}_N \pm \sqrt{\mathrm{D}\bar{k}/N}$
0.1	-0.00018 ± 0.00005	0.9987 ± 0.0005
0.3	-0.00115 ± 0.00007	0.9918 ± 0.0005
0.5	-0.00278 ± 0.00008	0.9791 ± 0.0006

Table 3.10. Estimates for the inhomogeneous ball

ε	$\bar{\tau}_N \pm \sqrt{\mathrm{D}\bar{\tau}/N}$	$\bar{k}_N \pm \sqrt{\mathrm{D}\bar{k}/N}$
0.1	-0.00006 ± 0.00001	0.9997 ± 0.0001
0.3	-0.00016 ± 0.00001	0.9982 ± 0.0003
0.5	-0.00045 ± 0.00001	0.9962 ± 0.0003

Table 3.11. Estimates of variances for the inhomogeneous ball

ε	$\sqrt{\mathrm{D}\tau}$	$\sqrt{\mathrm{D}k}$
0.1	0.0036	0.0446
0.3	0.0114	0.1448
0.5	0.0196	0.2379

We modelled the trajectories for the values $\tau_0 = 0.0675$ and $k_0 = 1/0.39$, which ensure the finiteness of the variances of the successful estimates in the whole range of the values of ρ [50]. In order to estimate the value Ek we had to additionally model branching with coefficient $\nu_0 = 1.5$. The preliminary calculations show that the satisfactory results are obtained even for $m = 2$, or $m = 3$, and for $M = 1$ [see (3.41)]. In order to model the trajectories in the inhomogeneous ball we employed the 'maximum section' method described in Section 3.1. The estimates in Table 3.9 are obtained from $N = 10^6$ and $N = 4 \times 10^6$ sampling tragectories for $\tilde{\tau}_N$ and $\tilde{k}_N$, respectively. In the case of the inhomogeneous ball (see Table 3.10) $N = 32 \times 10^6$ but $N = 100 \times 10^6$ when we calculate $\tilde{k}_N$ for $\varepsilon = 0.1$.

In these tables $\tilde{\tau}$ and $\tilde{k}$ are the above randomized sample estimates for $M = 1$; $\tilde{\tau}_N$ and $\tilde{k}_N$ are the corresponding successful estimates for $\mathrm{E}\tau$ and $\mathrm{E}k$. The results given in Table 3.10 are obtained with $m = 3$ for $\tilde{\tau}_N$ and with $m = 2$ for $\tilde{k}_N$.

It is not difficult to see that the data in Tables 3.7-9 agree satisfactorily. Analysis of the estimates for $\mathrm{E}\tau$ and $\mathrm{E}k$ in Table 3.9 shows that they are rather close to the corresponding values of the average density.

Besides, we have obtained the estimates of the values $\mathrm{D}\tau$ and $\mathrm{D}k$ for the two-layer medium with $m = 2$, $M = 1$, and $N = 10^7$ by the Monte Carlo method. They are presented in Table 3.11. Thus, due to the introduction of the second layer the values $\mathrm{D}\tau$, $\mathrm{D}k$ decrease but not as much as $|\mathrm{E}\tau|$ and $1 - \mathrm{E}k$.

We now consider the application of the averaging formulae (3.37) to the two-layer ball after the replacement of ρ by ρ_0 and k by $C\rho_0$, where $\rho_0 = (\rho_1 + \rho_2)/2$. A detailed analysis of the statistical estimates shows that $\mathrm{E}(k|\varepsilon = 0.1) \approx 0.9997$ is somewhat underestimated. Therefore, as the value of γ we use $\gamma_1' = 0.014$ for which $k_1(0.1) = 0.99982$ (see Section 3.4.5) differs from the above numerical estimate only by σ. The corresponding values

$$k_2(0.3) = 0.99827, \qquad k_3(0.5) = 0.99480$$

are in rather good agreement with the numerical estimates. It is not difficult to understand that this agreement must be more rigorous for a multilayer (i.e. small-scale) stochastic system.

3.4.4. Use of the simplest mathematical models

The one-speed process of particle transport in the bounded medium with isotropic scattering is defined by the probabilities σ_s/σ and σ_f/σ of scattering and fission, respectively, at the collision point (we put $\sigma = \sigma_s + \sigma_f$), the number, ν, of particles per fission event, and the probability $\mathrm{P}(\mathbf{r}, \rho)$ of a particle escaping from the system after it is scattered at the point $\mathbf{r}$, where ρ is the random density of the homogeneous bounded medium considered.

Let us proceed to the simple probabilistic model of particle transport by averaging the function $\mathrm{P}(\mathbf{r}, \rho)$ over the variable $\mathbf{r}$. We consider an infinite plane layer bounded with respect to the variable $z : 0 \leq z \leq R$. When the direction ω of a particle after scattering is fixed, for the case $(\omega, \mathbf{e}_z) < 0$ we have

$$\begin{aligned} P(z,\rho) &= \exp(-\rho\sigma C_\omega z), \quad 0 \leq z \leq R, \\ \int_0^R P(z,\rho)\mathrm{d}z &= \frac{1 - \exp(-\rho\sigma C_\omega R)}{\rho\sigma C_\omega} \asymp \frac{\gamma_\omega}{\rho} \end{aligned} \tag{3.43}$$

for sufficiently large values of R.

It is clear that the asymptotics (3.43) must also hold at least approximately for a sufficiently wide class of geometrical forms of large volume. Therefore we put for the simplest model:

$$P(\mathbf{r},\rho) \approx P_0(\rho) = \frac{\gamma}{\rho}.$$

In order to determine $k(\rho)$ we calculate the average number of collisions $S(\rho)$ of the particles in this system, which are obtained from the 'unit power' source, i.e. from a single particle after the replacement $\nu \to \nu/k$. It is obvious that the recursion relation

$$S(\rho) = \left(1 - P_0(\rho)\right)\left[1 + \left(\frac{\sigma_s}{\sigma} + \frac{\nu\sigma_f}{k\sigma}\right) S(\rho)\right],$$

holds, from which we obtain for the case $S(\rho) < +\infty$:

$$S(\rho) = \left(1 - P_0(\rho)\right)\left[1 - \frac{\sigma_s + \nu\sigma_f/k}{\sigma}\left(1 - P_0(\rho)\right)\right]^{-1} = \left(1 - P_0(\rho)\right)T_\rho^{-1}(k).$$

Another important functional is the 'system multiplication factor' $S_1(\rho)$, which is equal to the average number of particles that escape from the system from the 'unit power' source. In the simplest model we have the relation

$$S_1(\rho) = P_0(\rho) + \left(1 - P_0(\rho)\right)\left(\frac{\sigma_s}{\sigma} + \frac{\nu\sigma_f}{k\sigma}\right) S_1(\rho),$$

and if $S_1(\rho) < +\infty$, then $S_1(\rho) = P_0(\rho)T_\rho^{-1}(k)$.

The value $k(\rho) = k_0(\rho)$ is the critical value of the parameter k and is defined by the root of the equation $T_\rho(k) = 0$, i.e.

$$k_0(\rho) = \nu\sigma_f\left(1 - P_0(\rho)\right)[\sigma_f + P_0(\rho)\sigma_s]^{-1} = \nu\sigma_f(\rho - \gamma)(\sigma_f\rho + \gamma\sigma_s)^{-1}. \tag{3.44}$$

From the equation $k_0(1) = 1$, we have for the parameters $\sigma_s = 0.97$, $\sigma_f = 0.03$, $\nu = 2.5$, from Section 3.4.2:

$$\gamma = \frac{(\nu - 1)\sigma_f}{\sigma + (\nu - 1)\sigma_f} = 0.043062201.$$

Further we obtain for ρ, which is uniformly distributed in the interval $(1 - \varepsilon, 1 + \varepsilon)$, the formulae:

$$\mathrm{E}[k_0|\varepsilon] = \nu\left(1 - \frac{\gamma\sigma}{2\varepsilon\sigma_f}\ln\frac{\sigma_f(1+\varepsilon) + \gamma\sigma_s}{\sigma_f(1-\varepsilon) + \gamma\sigma_s}\right),$$

$$\mathrm{E}[k_0^2|\varepsilon] = \nu\left(-\nu + 2\mathrm{E}k_0 + \frac{\nu\gamma^2}{(\sigma_f + \gamma\sigma_s)^2 - (\sigma_f\varepsilon)^2}\right).$$

The results of calculations by these formulae are given in Table 3.12.

The values of $\mathrm{E}k_0$ in Table 3.12 are sufficiently close to the corresponding values in Table 3.8, and the values of the mean square deviations $(\mathrm{D}k_0)^{1/2}$ are about 40 percent less. This is due to the decreased variance because of averaging of the probability of a particle escaping.

From the expression

$$\frac{k_0(\rho)}{\nu} = 1 - \frac{\gamma\sigma}{\sigma_f\rho + \gamma\sigma_s},$$

it follows that $k_0'(\rho) > 0$ and $k_0''(\rho) < 0$. Hence for the simplest model we have $\mathrm{E}k_0(\rho) \le k_0(\mathrm{E}\rho)$ by virtue of the familiar Jensen inequality. However, the computational results in

Sections 3.4.2, 3.4.3 which are obtained by the more exact models, show that the converse relation $\mathrm{E}k > k(\mathrm{E}\rho)$ can hold for the system of relatively small volume. It turns out that this relation may be obtained in the framework of a simple model if, as in (3.43), we use the approximation for the average escape probability

$$P(\mathbf{r},\rho) \approx P_1(\rho) = \frac{\beta}{\rho}(1 - e^{-\rho\sigma R}).$$

Table 3.12. Estimates of $\mathrm{E}k_0$, $\sqrt{\mathrm{D}k_0}$

ε	$\mathrm{E}k_0$	$\sqrt{\mathrm{D}k_0}$
0.1	0.9991	0.0362
0.3	0.9921	0.1099
0.5	0.9776	0.1870

Note that unlike $P_0(\rho)$ the function $P_1(\rho)$ is not concave. For the modified model considered we have

$$k(\rho) = k_1(\rho) = \frac{\nu\sigma_f[\rho - \beta(1 - \exp(-\rho\sigma R))]}{\rho\sigma_f + \sigma_s\beta(1 - \exp(-\rho\sigma R))}.$$

For $R = 3$ and $\beta = 0.18836439$ we have $k_1(1) = 0.30241150$, which practically coincides with the diffusion approximation of the value $k(1)$ for the corresponding version of the spherical problem in Section 3.4.2. The direct calculations lead to the results:

$$\mathrm{E}[k_1|0.3] - k_1(1) = 0.0000001, \quad \mathrm{E}[k_1|0.5] - k_1(1) = 0.000064,$$

i.e. in this case we have $\mathrm{E}k_1 > k_1(\mathrm{E}\rho)$ for $\varepsilon \geq 0.3$.

We find rather easily the critical value k^* of the parameter k for the averaged solution (see the end of Section 3.4.1) in the framework of the simplest model. The expression is true for ρ, which is uniformly distributed in the interval $(1 - \varepsilon, 1 + \varepsilon)$:

$$\begin{aligned} R(1/k;\varepsilon) &= \mathrm{E}\left[\frac{S_1(\rho)}{\sigma}\right] \\ &= \frac{\gamma k}{\sigma_f(k-\nu)2\varepsilon}\ln\frac{[(1+\varepsilon)\sigma_f+\gamma\sigma_s]k-\nu\sigma_f(1+\varepsilon-\gamma)}{[(1-\varepsilon)\sigma_f+\gamma\sigma_s]k-\nu\sigma_f(1-\varepsilon-\gamma)}. \end{aligned} \tag{3.45}$$

It is obvious that the inequality $k_0(\rho) < \nu$ holds and $k_0(\rho)$ increases monotonically with ρ in the probabilistic model considered. Hence the value R in (3.45) is finite for $k_0(1+\varepsilon) < k < \nu$. It is not difficult to see that the numerator of the fraction under the logarithm sign in (3.45) vanishes for $k = k_0(1+\varepsilon)$. This implies that $k^*(\varepsilon) = k_0(1+\varepsilon)$ as noted at the end of Section 3.4.1.

We consider the possibility of estimating the value $k^*(\varepsilon)$ in terms of the radius of convergence of the power series

$$R(\lambda;\varepsilon) = \frac{C}{1/\nu - \lambda}\ln\frac{\sigma\gamma - c(1/\nu - \lambda)}{\sigma\gamma - d(1/\nu - \lambda)} = C\sum_{n=0}^{\infty} a_n\left(\lambda - \frac{1}{\nu}\right)^n, \quad \lambda = \frac{1}{k}.$$

This estimate may be useful when the conditions of Proposition 1.1 (see Section 3.4.1) are not satisfied. Here $c = -\nu\sigma_f(1+\varepsilon-\gamma)$, $d = -\nu\sigma_f(1-\varepsilon-\gamma)$. We choose $\lambda_0 = 1/\nu$

because particularly simple expressions for the values a_n are obtained in this case. Besides, there is no branching in the corresponding algorithm of the Monte Carlo method, and the variances of the estimates are obviously finite. The following expression turns out to be true:

$$\begin{aligned} R(\lambda;\varepsilon) &= \frac{C}{1/\nu-\lambda}\ln\frac{1-c(1/\nu-\lambda)/(\sigma\gamma)}{1-d(1/\nu-\lambda)/(\sigma\gamma)} \\ &= C\sum_{n=0}^{\infty}\frac{1}{n+1}\left[\left(\frac{d}{\sigma\gamma}\right)^{n+1}-\left(\frac{c}{\sigma\gamma}\right)^{n+1}\right]\left(\frac{1}{\nu}-\lambda\right)^{n}. \end{aligned}$$

From this expression follows the relation

$$\left|\frac{a_{n-1}}{a_n}\right| = \frac{\sigma\gamma}{|c|}\left(1+\frac{1}{n}\right)\left(1+\alpha^n\frac{\alpha-1}{1-\alpha^{n+1}}\right) = \left(1+\frac{1}{n}\right)A_n, \tag{3.46}$$

where

$$\alpha = \frac{d}{c} = \frac{1-\varepsilon-\gamma}{1+\varepsilon-\gamma}.$$

Thus, $|a_{n-1}/a_n| \to \sigma\gamma/|c|$ and

$$\lambda_0 - \frac{\sigma\gamma}{c} = \frac{1}{\nu} - \frac{\sigma\gamma}{c} = \frac{1}{k_0(1+\varepsilon)} = \frac{1}{k^*(\varepsilon)}.$$

However, the convergence in (3.46) is very slow since it is defined by the multiplier $1+1/n$. Hence we can try to improve the estimate of the radius of convergence in the actual problem, using the value $(na_{n-1})/[(n+1)a_n]$ rather than a_{n-1}/a_n. In particular, we have $A_2 = 0.9982$ and $A_3 = 0.9999$ for $\varepsilon = 0.5$.

Let us consider the possibilities of estimating the value τ in the framework of the simplest probabilistic particle transport model. Putting

$$\sigma_s + \sigma_f = \sigma, \qquad P(\mathbf{r},\rho) \approx P_0(\sigma\rho+\tau) = \frac{\gamma}{\sigma\rho+\tau}$$

in accordance with (3.44) we have the equation for $\tau(\rho)$:

$$\nu\rho\sigma_f\left(1-\frac{\gamma}{\sigma\rho+\tau}\right)\left[\tau+\sigma_f\rho+\frac{\gamma\rho\sigma_s}{\sigma\rho+\tau}\right]^{-1} = 1.$$

However, this equation turns out to yield the values of $\tau(\rho)$, which are substantially underestimated in magnitude, the values of $|\mathrm{E}[\tau|\varepsilon]|$ being underestimated by a factor of 10. We can obtain a better result if we put $\tau\varphi \approx \rho\tau\hat{\varphi}$, add the value $-\tau\rho\hat{\varphi}$ and $\nu\sigma_f\rho\hat{\varphi}$, and represent the obtained equation as (3.29), replacing $\nu \to (\nu\sigma_f - \tau)/\sigma_f$. When $P(\mathbf{r},\rho) \approx P_0(\rho) = \gamma/\rho$, the corresponding equation for $\tau(\rho)$ is

$$\frac{(\nu\sigma_f-\tau)(\rho-\gamma)}{\sigma_f\rho+\gamma\sigma_s} = 1. \tag{3.47}$$

This equation yields the expressions

$$\tau_0(\rho) = \nu\sigma_f - \frac{\sigma_f\rho + \gamma\sigma_s}{\rho - \gamma}, \quad \mathrm{E}(\tau_0|\varepsilon) = \sigma_f\left[\nu - 1 - \frac{\gamma}{2\varepsilon\sigma_f}\ln\frac{1+\varepsilon-\gamma}{1-\varepsilon-\gamma}\right],$$

$$\mathrm{E}(\tau_0^2|\varepsilon) = (2\mathrm{E}\tau_0 - \nu + 1)\sigma_f(\nu - 1) + \frac{\gamma^2}{(1-\gamma)^2 - \varepsilon^2}.$$

The corresponding estimates for the problem in Section 3.4.2 for $\gamma = 0.043062201$ are given in Table 3.13.

In spite of the simplicity of the model used, these estimates are in rather good agreement with the results of the more exact calculations in Sections 3.4.2, 3.4.3. The inequalities $\tau_0' > 0$, $\tau_0'' < 0$, which confirm the relation $\mathrm{E}(\tau_0) < \tau_0(\mathrm{E}\rho)$, hold. It is interesting to note that this relation also holds in the framework of the modified model with

$$P(\mathbf{r},\rho) \approx P_1(\rho) = \frac{\beta(1 - e^{-\rho\sigma R})}{\rho}$$

where the value $\beta = 0.18836439$ corresponds to $R = 3$, i.e. the relation holds for the version in which $\mathrm{E}(k) > k(\mathrm{E}\rho)$ for $\varepsilon > \varepsilon_0 > 0$ as indicated above. The expression for the value $\tau(\rho)$, which is obtained from (3.47) after the replacement $\gamma \to \beta(1 - e^{-\rho\sigma R})$, yields the values for $\varepsilon = 0.5$:

$$\mathrm{E}[\tau(\rho)|0.5] \approx -0.192, \qquad \tau(\mathrm{E}\rho) = -0.173.$$

These values practically coincide with the corresponding values obtained for $R = 3$ in the more exact diffusion approximation.

We consider the estimate of the critical value τ^* of the parameter τ for the averaged solution in terms of the radius of convergence of the appropriate power series. Substituting the values $k = 1$ and replacing ν by $(\nu\sigma_f - \tau)/\sigma_f$ in the expression $R_1(\tau;\varepsilon) = \mathrm{E}[S_1(\rho)/\sigma]$, by analogy with the above we get

$$R_1(\tau;\varepsilon) = C_1\sum_{n=0}^{\infty}\frac{1}{n+1}\left[\left(\frac{c_1}{\sigma\gamma}\right)^{n+1} - \left(\frac{d_1}{\sigma\gamma}\right)^{n+1}\right](\tau^{(0)} - \tau)^n,$$

where

$$\tau^{(0)} = (\nu - 1)\sigma_f, \quad d_1 = -(1 - \varepsilon - \gamma), \quad c_1 = -(1 + \varepsilon - \gamma).$$

Note that the value $\tau^{(0)} = (\nu - 1)\sigma_f$ corresponds to $\lambda_0 = 1/\nu$ in that these values are critical for the version of the simplest model with $P_0 \equiv 0$. Thus, relation (3.46) and all of its corollaries hold. In particular,

$$\tau^{(0)} + \lim|a_{n-1}/a_n| = \tau_0(1+\varepsilon) = \tau^*(\varepsilon),$$

since we have $R_1(\tau_0(1+\varepsilon);\varepsilon) = +\infty$ by analogy with the above, i.e. the value $\tau_0(1+\varepsilon)$ is actually the critical value of the parameter τ for the averaged solution.

Further we consider a simple method of estimating the critical parameters for a model system that is divided into n parts of equal volume with independent values of the density ρ chosen from the uniform distribution in $(1-\varepsilon, 1+\varepsilon)$. In this case $\mathrm{E}\rho = 1$, $\mathrm{E}\rho^2 = \varepsilon^2/(3n)+1$ and with allowance for relation (3.35), formulae (3.37) for $k \approx C\rho$, $\sigma_f^{(0)} \approx \mathrm{E}\rho_0 = \sigma_f$ have the form

Table 3.13. Estimates of $E\tau_0$, $\sqrt{D\tau_0}$

ε	$E\tau_0$	$\sqrt{D\tau_0}$
0.1	−0.00016	0.00274
0.3	−0.0016	0.0088
0.5	−0.0049	0.0171

$$\sigma^{(0)} = \sigma\left(1 + \frac{\varepsilon^2}{3n(1-\gamma)}\right) = \sigma\rho', \quad \sigma_s^{(0)} = \sigma_s\left(1 + \frac{\varepsilon^2}{3n}\right), \quad \sigma_f^{(0)} = \sigma_f.$$

On simple transformations the corresponding equation of the form (3.30) can be represented as

$$(\omega, \text{grad}\varphi) + \sigma\rho'\varphi - \sigma_s\rho'\hat{\varphi} = \frac{\nu\sigma_f\rho'}{\rho'}\left[1 - \frac{\varepsilon^2\sigma_s\gamma}{\nu\sigma_f 3n(1-\gamma)}\right]\hat{\varphi}.$$

This implies that as the estimate of the value of Ek we can use the value

$$\tilde{k}(\varepsilon) = \frac{k(\rho')}{\rho'}\left[1 - \frac{\varepsilon^2\sigma_s\gamma}{\nu\sigma_f 3n(1-\gamma)}\right]. \tag{3.48}$$

In Sections 3.4.2, 3.4.3 we give the estimates $k_2 = \tilde{k}(0.3)$ and $k_3 = \tilde{k}(0.5)$. The value γ is chosen so that the value of $\tilde{k}(0.1)$ coincides with the corresponding numerical estimate. The value $k(\rho')/\rho'$ is determined sufficiently accurately by quadratic interpolation, using the diffusion estimates $k(1.025) \approx 1.026645$, $k(1.05) \approx 1.052926$ and $k(1) = 1$:

$$k(\rho')/\rho' \approx 1 + (\rho' - 1)[0.064195 - 0.338472(\rho' - 1.025)].$$

For the model system under study, formulae (3.33) take the form

$$\sigma^{(0)} = \sigma, \quad \sigma_{s,f}^{(0)} = \rho'\sigma_{s,f}, \tag{3.49}$$

where

$$\rho' = \frac{E(\rho^2/(\rho-\gamma))}{E(\rho/(\rho-\gamma))}.$$

In accordance with (3.36) we have for $n = 1$:

$$\rho' = \rho'(\varepsilon) = \frac{1 + \gamma + \frac{\gamma^2}{2\varepsilon}\ln\frac{1+\varepsilon-\gamma}{1-\varepsilon-\gamma}}{1 + \frac{\gamma}{2\varepsilon}\ln\frac{1+\varepsilon-\gamma}{1-\varepsilon-\gamma}}.$$

On simple transformations the corresponding equation (3.30) reduces to the form

$$(\omega, \text{grad}\varphi) + \sigma\varphi - \sigma_s\hat{\varphi} = \nu\sigma_f[\rho' + \sigma_s(\rho' - 1)/(\nu\sigma_f)]\hat{\varphi}.$$

The critical value of the parameter k is equal to

$$k'(\varepsilon) = \rho' + \sigma_s(\rho' - 1)/(\nu\sigma_f),$$

By virtue of the known relation between the eigenvalues of k and τ for one-speed isotropic scattering [77], we can use the value

$$\tilde{\tau}(\varepsilon) = \nu\sigma_f \frac{k'(\varepsilon) - 1}{k'(\varepsilon)} = \frac{(\nu\sigma_f + \sigma_s)[\rho'(\varepsilon) - 1]}{1 + \sigma_s[\rho'(\varepsilon) - 1]/(\nu\sigma_f)}. \tag{3.50}$$

as the estimate of the value $\mathrm{E}\tau$.

The above values τ_2 and τ_3 are calculated by formula (3.50) by analogy with the values k_2 and k_3.

Note that the value $k(\rho')/\rho'$ varies rather weakly. Therefore formula (3.48) shows the variation of the value $1 - k(\varepsilon)$ with ε, which is confirmed by the computational results (see Sections 3.4.2, 3.4.3). Moreover, we see from formula (3.48) that the similarity rule holds, viz. the value $(1 - k(\varepsilon))\varepsilon^{-2}$ for different values of ε is approximately equally dependent on n (evidently stronger than $1/n$ since the value of γ for $n = 2$ is nearly twice as little as that for $n = 1$). This rule is also consistent with the computational results.

It is clear that the accuracy of the averaging procedure considered increases as the correlation scale of the random medium decreases. The results in Section 3.4.3 show that this averaging may be rather satisfactory for the two-layer medium. However, we may try to improve the procedure for the two-layer ball if, in formulae (3.34), instead of the approximation (3.35) we use the relations

$$\varphi_0^*(r, \omega) \approx C_0 \left(1 - \frac{\gamma}{\rho_1}\right) \hat{\varphi}_0^*(r), \qquad r < R \cdot 2^{-1/3}$$

$$\varphi_0^*(r, \omega) \approx C_0 \left(1 - \frac{\gamma^2}{\rho_1 \rho_2}\right) \hat{\varphi}_0^*(r), \qquad r > R \cdot 2^{-1/3}$$

and then put $\hat{\varphi}_0^*(r) \approx \text{const}$, $k \approx C\rho_0 = C(\rho_1 + \rho_2)/2$. We assume here that the escape probability from the outer layer is defined by the value of ρ_1 and from the inner layer by the product of the probabilities γ/ρ_1 and γ/ρ_2. Thus, formulae (3.34) reduce to the form

$$\sigma^{(0)} = \sigma \frac{2(\frac{\varepsilon^2}{3} + 2 - \gamma) - \gamma^2(1 + \frac{1}{2\varepsilon}\ln\frac{1+\varepsilon}{1-\varepsilon})}{4 - \gamma - \frac{\gamma}{2\varepsilon}(1 + 2\gamma)\ln\frac{1+\varepsilon}{1-\varepsilon}} = \sigma\rho',$$

$$\sigma_s^{(0)} = \sigma_s \left(1 + \frac{\varepsilon^2}{6}\right), \quad \sigma_f^{(0)} = \sigma_f,$$

$$\tilde{k}(\varepsilon) = \frac{k(\rho')}{\rho'}\left[1 + \frac{\sigma_s}{\nu\sigma_f}\left(1 + \frac{\varepsilon^2}{6} - \rho'\right)\right].$$

The corresponding calculations for $\gamma = 0.014$ yield the results:

$$\tilde{k}(0.1) = 0.99981, \qquad \tilde{k}(0.3) = 0.99815, \qquad \tilde{k}(0.5) = 0.99384$$

i.e. the new escape model markedly improves the value of $\tilde{k}(0.3)$ and somewhat deteriorates the value of $\tilde{k}(0.5)$ (see Section 3.4.3). It should be noted that $\varepsilon = 0.5$ implies too strong (up to 50 percent) fluctuation of the density.

The averaging procedure may be efficient, in particular, because the value $\mathrm{E}k$ can be estimated for small values of ε by the perturbation theory or when calculating the derivatives of k with respect to $\{\sigma_i\}$ by the Monte Carlo method.

3.4.5. Use of the second order parametric derivatives

We consider a random medium with the macroscopic cross-sections

$$\tilde{\sigma} = \rho_j \sigma, \qquad \tilde{\sigma}_s = \rho_j \sigma_s, \qquad \tilde{\sigma}_f = \rho_j \sigma_f,$$

where σ, σ_s, and σ_f are constants, and ρ_j is a positive random value. As a nonhomogeneous medium we consider the medium that consists of several homogeneous domains distinguished by the subscript j). The values ρ_j are independent.

In this Section we calculate $\mathrm{E}\tau^*$ and $\mathrm{E}k_{\text{eff}}$ by a method based on the following Taylor expansion in ρ_j in a neighbourhood of ρ_0 [38]:

$$k_{\text{eff}}(\rho) = k(\rho_0) + \sum_j k'_{\rho_j}(\rho_0)(\rho_j - \rho_0) + \sum_j \sum_l \frac{k''_{\rho_j \rho_l}(\rho_0)}{2}(\rho_j - \rho_0)(\rho_l - \rho_0) + \dots .$$

Here $k \equiv k_{\text{eff}}$. Using this equality, we obtain

$$\mathrm{E}k_{\text{eff}}(\rho) \approx k(\rho_0) + \sum_j k'_{\rho_j}(\rho_0)(\mathrm{E}\rho - \rho_0) + \sum_j \frac{k''_{\rho_j}(\rho_0)}{2}\mathrm{E}(\rho_j - \rho_0)^2.$$

An analogous expression can be obtained for $\mathrm{E}\tau^*(\rho)$.

We use the special iterative method for estimating k'_{eff}, k''_{eff}, $\tau^{*\prime}$, and $\tau^{*\prime\prime}$. The method is based on the following relations (see Section 3.2):

$$\lim_{m\to\infty} \frac{mJ_1^{(m-1)}(\lambda_0)}{J_1^{(m)}(\lambda_0)} = \lambda^* - \lambda_0 \quad \text{and} \quad \lim_{m\to\infty} \frac{mJ_2^{(m-1)}(\tau_0)}{J_2^{(m)}(\tau_0)} = \tau^* - \tau_0. \tag{3.51}$$

Here τ_0 and λ_0 are some values of τ and λ used in calculations, $J_1(\tau)$ and $J_2(\lambda)$ are linear functionals of the flux densities φ_1 and φ_2, and $J_1^{(m)}(\lambda_0)$ and $J_2^{(m)}(\tau_0)$ are derivatives of the functionals with respect to parameters λ and τ at the points $\lambda = \lambda_0$ and $\tau = \tau_0$.

Provided that in (3.51) the passage to the limit and the parametric differentiation are commutative, we deduce

$$k''_{\text{eff}} \approx m \frac{\left(J_1^{(m)}\right)''_\rho J_1^{(m-1)} - J_1^{(m)}\left(J_1^{(m-1)}\right)''_\rho}{\left(mJ_1^{(m-1)} + \lambda J_1^{(m)}\right)^2}$$
$$- 2m \frac{\left[\left(J_1^{(m)}\right)'_\rho J_1^{(m-1)} - J_1^{(m)}\left(J_1^{(m-1)}\right)'_\rho\right]\left[m\left(J_1^{(m-1)}\right)'_\rho + \lambda\left(J_1^{(m)}\right)'_\rho\right]}{\left(mJ_1^{(m-1)} + \lambda J_1^{(m)}\right)^3};$$

$$\tau^{*\prime\prime} \approx m \frac{\left(J_2^{(m-1)}\right)''_\rho J_2^{(m)} - J_2^{(m-1)}\left(J_2^{(m)}\right)''_\rho}{\left(J_2^{(m)}\right)^2}$$
$$- 2m \frac{\left[\left(J_2^{(m-1)}\right)'_\rho J_2^{(m)} - J_2^{(m-1)}\left(J_2^{(m)}\right)'_\rho\right]\left(J_2^{(m-1)}\right)'_\rho}{\left(J_2^{(m)}\right)^3}.$$

We calculate the functionals $J_\alpha = (\tilde{\sigma}\varphi_\alpha, h_\alpha)$ and their derivatives by the Monte Carlo method using the collision estimate (see Section 1.3):

$$\xi_\alpha = \sum_{n=1}^{N} Q_n h_\alpha(\mathbf{x}_n).$$

Here $E\xi_\alpha = J_\alpha$ $(\alpha = 1, 2)$, N is the number of collisions the particle make before escape or capture, $\mathbf{x}_n$ are the points where the collisions occur $(n = 1, \dots, N)$. Constructing the weights Q_n is described in Section 3.2. The direct differentiating the weights Q_n yields the weights for the derivatives of the functionals J_α (Section 3.2).

Let us take the functional J_2 of the form $J_2 = (\bar{\sigma}\varphi_2, \tilde{h}_2\bar{\sigma}_\theta(\rho_0)/\bar{\sigma}_\theta(\rho))$, where θ stands for s, f, or c, and the function $\tilde{h}_2$ does not depend on τ. This functional yields simple (exponential) weights. To obtain an estimate of the derivatives, we differentiate the weights $\tilde{Q}_n = Q_n\bar{\sigma}_\theta(\rho_0)/\bar{\sigma}_\theta(\rho)$. When calculating J_2 we use the substitution

$$\bar{\sigma}_f \to \bar{\sigma}_f + \rho\sigma_c + \frac{\tau}{v}, \qquad \nu \to \frac{\nu\bar{\sigma}_f}{\bar{\sigma}_f + \rho\sigma_c + \tau/v} \le \nu$$

that diminishes the statistical error (see Section 3.1).

As an example, we consider the homogeneous ball and the two-layer ball described in the previous section. While calculating, we use the functions $h_1 = \tilde{h}_2 = \sin(æ r)/r$, $æ \approx 0.3739866031$. They are approximately equal to the eigenfunction of the corresponding homogeneous problems.

Let us simulate $M = M_1 \cdot M_2$ trajectories. For every group consisting of M_1 trajectories, we build the estimate of the second derivative ζ_{m_2} and calculate the arithmetic mean $\zeta = M_2^{-1}\sum_{m_2=1}^{M_2}\zeta_{m_2}$. We obtain the corresponding mean square error by the formula $\sqrt{\left(M_2^{-1}\sum_{m_2}^{M_2}\zeta_{m_2}^2 - \zeta^2\right)/M_2}$. The value $m = 2$ was used.

Calculated derivatives and expectations of k_{eff} and τ^* for $M_1 = 10^5$ and $M_2 = 100$ are presented in Table 3.14. Here $\mathcal{J} = 1$ and $\mathcal{J} = 2$ correspond to the homogeneous ball and to the two-layer ball, respectively. The estimates of Ek_{eff} and $E\tau^*$ (Table 3.14) show that Ek_{eff} and $E\tau^*$ are close to the corresponding values for homogeneous averaged medium.

Table 3.14. Estimates obtained by the special iterative method

$\mathcal{J}$	k''_{ρ_1}	k''_{ρ_2}	Ek_{eff}, $\varepsilon = 0.1$
1	-0.578 ± 0.017		0.99904
2	-0.196 ± 0.016	0.064 ± 0.002	0.99978
$\mathcal{J}$	$\tau^{*''}_{\rho_1}$	$\tau^{*''}_{\rho_2}$	$E\tau^*$, $\varepsilon = 0.1$
1	-0.075 ± 0.004		-0.00013
2	-0.015 ± 0.003	-0.0031 ± 0.0003	-0.00003

3.5. New approach to path estimates in the Monte Carlo method

New weight modifications of the weight path estimate are constructed and substantiated. It seems, the new approach allows the path estimate to be extended for solving other kinds of integral equations.

1. The mathematical model of the transfer process is regarded as the uniform terminating Markov chain $x_0, x_1 \dots, x_N$ of particle collisions in the phase coordinate-velocity

phase $X = R \times V$, i.e. $x_n = (\mathbf{r}_n, \mathbf{v}_n)$, where $\mathbf{r}_n$ is the collision point and $\mathbf{v}_n$ is the velocity of the collided particle. This Markov chain is determined by the initial density $f_0(x)$ and the transition density $k(x', x)$ for the transition x' to x. It is supposed that

$$\int k(x', x)\mathrm{d}x = q(x') \leq 1 - \delta,\ \delta > 0,$$

i.e. the chain is termined with probability unity and the mean number of collisions is finite. For instance, this condition holds for bounded medium convexly surrounded by an auxiliary absorber. The kernel $k(x, x')$ is determined by the extinction coefficient

$$\sigma(x) = \sigma_s(x) + \sigma_c(x),$$

where $\sigma_s(x)$ is the coefficient of scattering with the given indicatrix $w(\mathbf{v}' \to \mathbf{v}; \mathbf{r})$, and $\sigma_c(x)$ is the absorption coefficient. The ratios $\sigma_s(x)/\sigma(x)$ and $\sigma_c(x)/\sigma(x)$ are equal to the probabilities of scattering and absorption immediately after a collision in the phase point x, and density distribution of the length l of the free path from r' to r is equal to $p(l) = \sigma(\mathbf{r}(l)) \exp(-\tau(l))$, where $\tau(l)$ is the optical length of the path (see Section 3.1).

The collision density $f(x)$ is the Neumann series for the integral equation of second kind is

$$f = Kf + f_0,$$

where K is an integral operator with the kernel $k(\cdot, \cdot)$. Although the kernel includes the generalized factor $\delta(\mathbf{v}/|\mathbf{v}| - (\mathbf{r} - \mathbf{r}')/|\mathbf{r} - \mathbf{r}'|)$, the operator K may be considered as acting from $L_1(X)$ to $L_1(X)$ (the more so here since all functions are non-negative). If $q(x') \geq 1 - \delta$, $\delta > 0$, then $\|K\|_{L_1} < 1$ and therefore $\rho(K) < 1$.

The radiation intensity $\Phi(x)$ is connected with the collision density by the relation: $f(x) = \sigma(x)\Phi(x)$, and (Φ, σ_s), (Φ, σ_c) are corresponding mean numbers of scatterings and captures for the given source. The Monte Carlo method is usually used for estimating the linear functionals:

$$J_\varphi = (f, \varphi) = (\sigma\Phi, \varphi).$$

Here we suppose that $\varphi \geq 0$ and $\varphi \in L_\infty(X)$. If $\{x_n\}_{n=0,1,\ldots,N}$ is the physical collision chain then

$$J_\varphi = \mathrm{E}\sum_{n=0}^{N} \varphi(x_n).$$

But one may use another Markov chain with the transition density $p(x', x)$ and the initial density $\pi(x)$, including the same generalized factors. There are introduced the auxiliary weights:

$$Q_0(x_0) = f_0(x_0)/\pi(x_0), \quad Q_n = Q_{n-1}k(x_{n-1}, x_n)/p(x_{n-1}, x_n).$$

If $supp\,\pi(\cdot) \supset supp\,f_0(\cdot)$ and $supp\,p(\cdot, \cdot) \supset supp\,k(\cdot, \cdot)$, then

$$J_\varphi = \mathrm{E}\xi, \quad \text{where} \quad \xi = \sum_{n=0}^{N} Q_n\varphi(x_n))$$

(see Section 1.3). If, additionally, $\rho(K_p) < 1$, where K_p is the operator with kernel $k^2(x', x)/p(x', x)$, then $\mathrm{D}\xi < +\infty$ (see Section 1.4).

In addition to the collision estimate ξ there is a standard capture estimate which is evaluated in a capture point, i.e. with the probability $g(x) = 1 - \int p(x, x')dx$; it is defined as

$$\eta = Q_N \varphi(x_N)/g(x_N).$$

If $supp\, g \supset supp\, \varphi$, then $E\eta = J_\varphi$. Besides, if

$$\varphi/g < c < +\infty \quad \text{and} \quad \rho(K_p) < 1$$

then $D\eta < +\infty$ (see Section 1.4).

Now let us suppose that all the above functions are vise-smooth dependent of a vector parameter $\lambda = (\lambda_1, \dots, \lambda_m)$ and consider the question about the possibility of computing the partial derivatives from function $J_\varphi(\lambda)$ on the basis of differentiating the collision estimate $\xi(\lambda)$ under special additional conditions. We shall use the notations:

$$F^{(s)}(x, \lambda) = \partial^s F(x, \lambda)/\partial \lambda_1^{s_1} \dots \partial \lambda_m^{s_m}, \quad s = \sum_{i=1}^{m} s_i.$$

Theorem 3.1. *Let the conditions*

$$[Q_n(\lambda)\varphi(x_n, \lambda)]^{(s)} \geq 0, \quad [f_0^{(s)}(\cdot, \lambda)]^2/\pi(\cdot) \in L_1(X), \quad \varphi^{(s)}(\cdot, \lambda) \in L_\infty,$$

hold, and suppose the operator with the kernel $|k^{(s)}(\cdot, \cdot, \lambda)|$ *is bounded and* $\rho(K_p) < 1$. *Then*

$$J^{(s)}(\lambda) = E\xi^{(s)}(\lambda), \quad \textit{and} \quad D\xi^{(s)}(\lambda) < +\infty.$$

Proof. The first statement follows from the first condition of the Theorem. The second statement is the direct consequence of Theorem 1.14. □

2. The weight path estimate is constructed and substantiated here for the functionals $J_h = (\Phi, h)$, with $h \geq 0$ и $h \in L_\infty$, by the special parametric differentiation. Firstly, an artificial additional capture is introduced with the coefficient $\sigma_c h(x)$, $\sigma_c \geq 0$, which is added to the initial physical capture coefficient $\sigma_c^{(0)}(r)$, i.e. the modified medium is considered with the capture coefficient

$$\sigma_c^{(h)}(x) = \sigma_c^{(0)}(x) + \sigma_c h(x), \quad \sigma_c = \text{const} \geq 0. \tag{3.52}$$

The corresponding radiation intensity is denoted by Φ_{σ_c}.

Lemma 3.1. *The following relation holds:*

$$(\Phi, h) = -\frac{dJ_{\sigma_c^{(0)}}(\sigma_c)}{d\sigma_c}|_{\sigma_c=0},$$

where

$$J_{\sigma_c^{(0)}}(\sigma_c) = (\Phi_{\sigma_c}, \sigma_c^{(0)}), \quad \Phi \equiv \Phi_{\sigma_c}|_{\sigma_c=0}. \tag{3.53}$$

Proof. The quantity $(\Phi_{\sigma_c}, \sigma_c^{(h)})$ is the total number of captures for the given source; therefore

$$(\Phi_{\sigma_c}, \sigma_c^{(h)}) = (\Phi_{\sigma_c}, \sigma_c^{(0)}) + (\Phi_{\sigma_c}, h\sigma_c) = C. \tag{3.54}$$

From (3.54) we have

$$-dJ_{\sigma_c^{(0)}}(\sigma_c)/d\sigma_c = (\Phi_{\sigma_c}, h) + \sigma_c d(\Phi_{\sigma_c}, h)/d\sigma_c.$$

The existence of the last derivative follows from Theorem 3.1 taking in account the kernel of the total fransfer equation (see Section 3.1).

Let us consider now Monte Carlo estimates for the quantity (Φ, h) on the basis of the expression (3.53). The functional $J_{\sigma_c^{(0)}}(\sigma_c)$ is the total number of physical captures (i.e. captures corresponding to the coefficient $\sigma_c^{(0)}$) for the given source. It is well known (see Section 3.2) that, if the direct simulation for the initial physical model is realized with $\sigma_c = 0$, then for the case with $\sigma_c \neq 0$ it is possible to use the weight estimate:

$$J_{\sigma_c^{(0)}}(\sigma_c) = \mathrm{E}q(L_N, \sigma_c), \quad q(L_N, \sigma_c) = \exp\{-\sigma_c \int_{L_N} h(x(l))\mathrm{d}l\}, \tag{3.55}$$

where L_N is the total particle path to a physical capture. So, in the case of the direct simulation it is possible to use the representation:

$$(\Phi, h) = \mathrm{E}\tau_h(L_N) = \mathrm{E}\int_{L_N} h(x(l))\mathrm{d}l. \tag{3.56}$$

The expression (3.56) was obtained in [24] earlier in a more complicated way for the case $h \equiv h(r)$ without investigating the variance finiteness. In [34], a weight modification of that path estimate was constructed with practically difficult integrating of a variable weight factor over a particle path. Our new approach enables us to construct a simplier weight modification, with which we can investigate the variance finiteness.

Theorem 3.2. *Let the following conditions*

$$g(x) \geq C_1 > 0 \ \ if \ \ x \in supp\, \sigma_c^{(0)}(\cdot), \ \sigma(x) \geq C_2 > 0 \ \ if \ \ x \in supp\, \sigma(\cdot)$$

hold and, besides, $\rho(K_p) < 1$ for the weight model of the initial problem. Then $(\Phi, h) = \mathrm{E}\zeta$, where

$$\zeta = \frac{Q_N \sigma_c^{(0)}(x_N)}{g(x_N)\sigma(x_N)} \tau_h(L_N) = Q_N^{(0)} \tau_h(L_N), \tag{3.57}$$

and $\mathrm{D}\zeta < +\infty$.

Proof. Let us represent functional $J_{\sigma_c^{(0)}}$ by the weighted 'capture estimate':

$$(\Phi_{\sigma_c}, \sigma_c^{(0)}) = (f_{\sigma_c}, \sigma_c^{(0)}/[\sigma + \sigma_c h])$$

$$= \mathrm{E}[Q_N(\sigma_c)\frac{\sigma_c^{(0)}(x_N)}{\sigma(x_N) + \sigma_c h(x_N)} \frac{1}{g(x_N)}] = \mathrm{E}[Q_N^{(0)} q(L_N, \sigma_c)].$$

The last equality follows directly from (3.55). Further, differentiating with respect to σ_c in the point $\sigma_c = 0$, we obtain (3.57).

To derive the relation $\mathrm{D}\zeta < +\infty$, we consider the standard collision estimate for the functional $(f_{\sigma_c}, \sigma_c^{(0)}/[\sigma + \sigma_c h])$:

$$\xi = \sum_{n=0}^{N} Q_n q(L_n, \sigma_c)\sigma_c^{(0)}(x_n)/\sigma(x_n).$$

It is not difficult to state that, if conditions of Theorem 3.2 hold, then conditions of Theorem 3.1 hold so. Therefore $(\Phi, h) = \mathrm{E}\zeta_1$, where

$$\zeta_1 = \sum_{n=0}^{N} Q_n \tau_h(L_n)\sigma_c^{(0)}(x_n)/\sigma(x_n), \tag{3.58}$$

and $\mathrm{D}\zeta_1 < +\infty$. Finally, it is enough to take into account that if the first condition of the Theorem holds then $\zeta \leq \zeta_1/C_1$. □

Note that, in [49] for the standard path estimate (i.e. for $h \equiv 1$), in fact the following weight modification is considered :

$$(\Phi, \delta_D) = \mathrm{E}\sum_{n=0}^{N} Q_n h_D(l_n),$$

where l_n is the path from r_{n-1} to r_n, δ_D indicates the region D, $h_D(l_n)$ is the length of that part of the path l_n, which belongs to D. The finiteness of the corresponding estimate variance is proved in [49] under the additional condition of the region D boundness. This approach provides the construction and substantiation of the more general estimate.

Theorem 3.3. *Under the conditions of Theorem 3.2 for the bounded supp h the value*

$$\zeta_2 = \sum_{n=0}^{N} Q_n \tau_h(l_n)$$

is the nonbiased estimate (Φ, h) with finite variance.

Proof. This is constructed by the special point-wise averaging of the value ζ from (3.57). Because of the relations $Q_N^{(0)} \geq 0$, $\tau_h(l_n) \geq 0$ the following equality for the random variables ζ from (3.57) holds:

$$\mathrm{E}\zeta = \mathrm{E}\sum_{n=0}^{N} Q_N^{(0)}\tau_h(l_n) = \sum_{n=0}^{\infty} \mathrm{E}\left[\Delta_n Q_N^{(0)}\tau_h(l_n)\right],$$

where Δ_n is the indicator of non-termination of a trajectory up to the state x_n (see Section 1.3). Further,

$$\mathrm{E}\left[\Delta_n Q_N^{(0)}\tau_h(l_n)|N \geq 0; x_0, ..., x_n\right] = Q_n\tau_h(l_n),$$

because $Q_N^{(0)}$ is a non-biased estimate of the capture probability and

$$\mathrm{E}(Q_N^{(0)}/Q_n|N \geq n) = 1.$$

Therefore,

$$\mathrm{E}\zeta_2 = \mathrm{E}\zeta = (\Phi, h).$$

In addition, under the theorem conditions the relation $\tau_h(l_n) < C < +\infty$ holds; hence $\zeta_2 \le \xi_2 = C\sum_{n=0}^{N} Q_n$, $D\xi_2 < +\infty$. □

It is interesting to note that in the proof of Theorem 3.2, another new weight modification (3.58) of the 'path estimate' is used which can be obtained from (3.57) by the special point-wise averaging. This averaging does not ensure the variance decreases, because it is realized for every point separately.

3. Now suppose that the vise-component sum of the vector $\{\sigma_c^{(i)} h_i(x)\} = H(x)$ is added to the initial capture coefficient $\sigma_c^{(0)}(x)$ in addition to the value $\sigma_c h(x)$ (here $\sigma_c^{(i)} \ge 0$ and $h_i(x) \ge 0$, $h_i \in L_\infty(X)$), $i = 1, \ldots, m$, i.e. instead of (1) the expression

$$\sigma_c^{(H)}(x) = \sigma_c^{(0)}(x) + \sigma_c h(x) + \sum_{i=1}^{m} \sigma_c^{(i)} h_i(x)$$

is considered. Under the conditions of Theorem 3.2 on the basis of the new weight modification of the path estimate the following representations for multiple derivatives in the point $\sigma_c^{(i)} = 0$, $i = 0, 1, \ldots, m$ are obtained

$$\frac{\partial^s(\Phi, h)}{\partial\sigma_c^{(1)s_1} \ldots \partial\sigma_c^{(m)s_m}} = \mathrm{E}[Q_N^{(0)} \tau_h(L_N) \prod_{i=1}^{m} \tau_{h_i}^{s_i}(L_N)], \quad \sum_{i=1}^{m} s_i = s. \tag{3.59}$$

Variances of the corresponding estimates are finite. Of some practical importance is the problem version with $h_i \equiv \delta_{X_i}$ for a partition $X = \bigcup_{i=1}^{m} X_i$. In this case, from (3.59), the following representations are obtained:

$$\frac{\partial(\Phi, \delta_{X_i})}{\partial\sigma_c^{(j)}} = \mathrm{E}[Q_N^{(0)} h_{D_i}(L_N) h_{D_j}(L_N)]. \tag{3.60}$$

In the case of the direct simulation, the expressions (3.60) represent the set of the second probabilistic moments of the random vector $\{h_{D_i}(L_N)\}$. So the matrix $\{\partial(\Phi, \delta_{X_i})/\partial\sigma_c^{(j)}\}$ is positively determined and possesses the property of the diagonal dominance. This indicates the high conformity of the values of $\{(\Phi, \delta_{X_i})\}$ with respect to the coefficients $\{\sigma_c^{(i)}\}$, $i = 1, \ldots, m$. The corresponding reverse problem is solvable by the Newton method by calculating functionals and their partial derivatives on the basis of the weight path estimate.

It is possible to construct similar representations using the estimate ζ_2 from Theorem 3.3 with $Q_n = Q_n\,(\sigma_c^{(1)}, \ldots, \sigma_c^{(m)})$. In this case

$$\frac{\partial^s(\Phi, h)}{\partial\sigma_c^{(1)s_1} \ldots \partial\sigma_c^{(m)s_m}} = \mathrm{E}\left[\sum_{n=0}^{N} Q_n \tau_n(l_n) \prod \tau_{h_i}^{s_i}(L_n)\right], \quad \sum_{i=1}^{m} s_i = s,$$

and, if $\operatorname{supp} h$ is bounded, then

$$\zeta^{(s)} \le C \sum_{n=0}^{N} Q_n \prod_{i=1}^{m} \tau_{h_i}^{s_i}(L_n) = C\xi_2^{(s)}.$$

Here ξ_2 is the collision estimate related to $\varphi \equiv C$, and $D\xi_2^{(s)} < +\infty$.

Note that by analogy with [55] it possible to prove that a path estimate can estimate the function

$$\Phi_h(r) = \int \Phi(x)h(x)\mathrm{d}v$$

essentially better then a collision estimate with respect to the metric $C(R)$. The corresponding advantage in computational cost is of the order of the factor δ, equal to the required error.

4. Let us consider the equation

$$\Phi(x) = \int_0^\infty q(x')k(x',x)\Phi(x')\mathrm{d}x' + e^{-x}, \tag{3.61}$$

where $0 \leq q(x) \leq 1$, $k(x',x) = \mathrm{E}r(x',x;\sigma)$, and $r(x',x;\sigma) = \sigma e^{-\sigma(x-x')}$ if $x \geq x'$, and $r \equiv 0$ if $x < x'$. Here σ is a random parameter with values from $[\sigma_0, +\infty)$, $\sigma_0 > 0$, $\mathrm{E}\sigma^{-1} < +\infty$.

Now let us introduce the auxiliary parameter $\sigma_c \geq 0$, and change the above exponent to $\exp\{-(\sigma+\sigma_c h(x'))(x-x')\}$. After integrating with respect to x instead of the modified equation (3.61) we obtain the relation:

$$(\Phi_{\sigma_c}, 1-q) + \sigma_c(\Phi_{\sigma_c}, \mathrm{E}[qh/(\sigma+\sigma_c h)]) = 1,$$

which provides constructing weight path estimates.

Equation (3.61) is the special generalization of the simplest integral transfer equation with δ-scattering. It is clear, that this approach can be extended to the analogous modifications of the total transfer equation (see Section 3.1).

Using the relation of (3.54) type, it is possible to extend the above results to continuous diffusion processes; here representation (3.56) holds also for linear functionals related to solutions of a stationary diffusion problem.

Finally note that, to extend the above statements for the case $h(\cdot) \not\geq 0$, it is sufficient to employ the general Theorem 1.14 and to derive relation (3.54) by the proper integration of the integral transfer equation, as in Item 4.

3.6. Monte Carlo estimates for derivatives of polarized radiation

Let us consider the equation of radiative transfer with polarization [49], [41] in $L_1(X)$:

$$\varphi_i(x) = \sum_{j=1}^4 \int_X K_{ij}(x',x)\varphi_j(x')\mathrm{d}x' + f_i(x), \quad i = 1,\dots,4, \tag{3.62}$$

or in the operator form $\Phi = \mathbf{K}\Phi + F$, and the equation, adjoint to it, in $L_\infty(X)$:

$$\varphi_i^*(x) = \sum_{j=1}^4 \int_X K_{ji}(x,x')\varphi_j^*(x')\mathrm{d}x' + h_i(x), \quad i = 1,\dots,4, \tag{3.63}$$

or in the operator form $\Phi^* = \mathbf{K}^*\Phi^* + H$.

Here $X = R \times \Omega$, where R is the space of coordinates, Ω is the space of directions. Norms in spaces of vector functions $L_1(X)$ and $L_\infty(X)$ are set by the equalities

$$\| F \|_{L_1(X)} = \sum_{i=1}^{4} \int_X |f_i(x)| \mathrm{d}x \quad \text{and} \quad \| H \|_{L_\infty(X)} = vraisup|h_i(x)|$$

accordingly.

The kernel of the operator **K** is represented as:

$$K(x', x) = \frac{\sigma(\mathbf{r}) \exp\{-\tau(\mathbf{r}', \mathbf{r})\} \sigma_s(\mathbf{r}') P(\omega', \omega, \mathbf{r}')}{\sigma(\mathbf{r}') |\mathbf{r}' - \mathbf{r}|^2} \delta(\omega - \mathbf{s}) \tag{3.64}$$

Here $\sigma_c(\mathbf{r})$ is the absorption coefficient , $\sigma_s(\mathbf{r})$ is the scattering coefficient, $\sigma(\mathbf{r}) = \sigma_c(\mathbf{r}) + \sigma_s(\mathbf{r})$ is the coefficient of full interaction with the medium; $\tau(\mathbf{r}', \mathbf{r}) = \int_0^{|\mathbf{r}'-\mathbf{r}|} \sigma(\mathbf{r}' + \mathbf{s}t) dt$ is the optical length of the segment $(\mathbf{r}', \mathbf{r})$; $\mathbf{s} = (\mathbf{r} - \mathbf{r}')/|\mathbf{r}' - \mathbf{r}|$, $P(\omega', \omega, \mathbf{r}')$ is the phase matrix, the vectors ω' and ω are directions before and after scattering at the point $\mathbf{r}'$, respectively [41].

The kernel $K(x', x)$ can be decomposed as

$$K(x', x) = l(\mathbf{r}', \mathbf{r}) * C(x', x),$$

where the matrix $C(x', x)$ is related to modifications of directions of driving particles at collisions, and the scalar function $l(\mathbf{r}', \mathbf{r})$ describes the transition of a particle from $\mathbf{r}'$ to $\mathbf{r}$. Thus, the operator **K** is a composition of the operators **l** and **C**:

$$K(x', x) = \int_X l_\delta(x'', x) C_\delta(x', x'') \mathrm{d}x'' \quad \text{or} \quad \mathbf{K} = \mathbf{lC},$$

where the kernel $l_\delta(x', x)$ of the transition operator **l** can be represented as:

$$l_\delta(x', x) = \frac{\sigma(\mathbf{r}) \exp\{-\tau(\mathbf{r}, \mathbf{r}')\}}{|\mathbf{r} - \mathbf{r}'|^2} \delta(\omega' - \mathbf{s}) \delta(\omega - \mathbf{s}),$$

and the kernel $C_\delta(x', x)$ of the scattering operator **C** can be represented as:

$$C_\delta(x', x) = \frac{\sigma_s(\mathbf{r}')}{\sigma(\mathbf{r}')} P(\omega', \omega, \mathbf{r}') \delta(\mathbf{r}' - \mathbf{r}).$$

The density function of the initial collisions $F(x)$ in eq.(3.62) is the Stokes vector function and can be expressed as $F(x) = [\mathbf{l}S](x)$, where

$$[\mathbf{l}S](x) = \int_X l_\delta(x', x) S(x') \mathrm{d}x',$$

$S(x)$ is the vector function of real radiation sources.

The vector collision density function $\Phi(x)$ is related to the classical Stokes vector function $I(x)$, describing an intensity, a degree of polarization, a plane of polarization and a degree of ellipticity of scattering radiation [41], as follows: $\Phi(x) = I(x)\sigma(\mathbf{r})$.

Let us consider the following linear functional:

$$\mathcal{J}_H = < \Phi, H > = < \Phi^*, F >,$$

where

$$<\Phi, H>=\int_X \Phi^T(x)H(x)\mathrm{d}x, \quad <\Phi^*, F>=\int_X F^T(x)\Phi^*(x)\mathrm{d}x.$$

For its calculation by the Monte-Carlo method we can construct the collision estimate ξ on the homogeneous Markov chain $x_0, \dots, x_N$ with the initial density function $\pi(x_0)$ and the transition density function $p(x', x) = p_1(\mathbf{r}', \mathbf{r})p_2(x', x)$, looking like:

$$\xi = \sum_{n=0}^{N} q_n^T H(x_n),$$

where the vector weight q_n are defined through the matrix weight Q_n by the recurrence formulae:

$$q_0 = \frac{F(x_0)}{\pi(x_0}, \quad Q_0 = \{\delta_{ij}\},$$

$$Q_n = Q_{n-1}\frac{K^T(x_{n-1}, x_n)}{p(x_{n-1}x_n)}, \quad q_n = Q_{n-1}^T q_0, \; n = 1, \dots, N. \tag{3.65}$$

It is easy to find that $\xi = q_0^T \xi_{x_0}$, where

$$\xi_{x_0} = \sum_{n=0}^{N} Q_n H(x_n). \tag{3.66}$$

Then, under appropriate conditions on the operator $\mathbf{K}$ (see below) the following equalities hold:

$$\mathcal{J}_H = \mathrm{E}\xi = \mathrm{E}_{x_0}\{q_0^T E(\xi_{x_0}|x_0)\}, \tag{3.67}$$

$$\mathrm{E}\xi_x = \Phi^*(x), \tag{3.68}$$

Hence the variable ξ_x can be considered as a local estimate of the solution Φ^* of adjoint transfer equation (3.63). For the validity of equality (3.68) it is necessary, generally speaking, to impose on the operator $\mathbf{K}$ the condition $\rho(\mathbf{K}_1) < 1$, where $\mathbf{K}_1$ is the integral operator obtained from $\mathbf{K}$ by the replacement of the components $K_{ij}(x', x)$ with their modules. Under the additional condition $\rho(\mathbf{K}_{1,p}) \leq q_p < 1$, the covariance matrix $\Psi(x) = \mathrm{E}(\xi_x[\xi_x]^T)$ is a solution of the equation

$$\Psi(x) = [H\Phi^{*T} + \Phi^* H^T - HH^T](x) + \int_X \frac{K^T(x, x')\Psi(x')K(x, x')}{p(x, x')}\mathrm{d}x', \tag{3.69}$$

or in operator form $\Psi = \chi + \mathbf{K}_p\Psi$ [49]. Hereinafter the index "1" will be used to denote an integral operator obtained from the initial one by the replacement of kernel components with their modules (for example $\mathbf{K}_1$), and the index "p" will denote an operator obtained from the initial one according to equation (3.69) (for example $\mathbf{K}_p$).

The variance of the scalar estimate ξ is related to the covariance matrix $\Psi(x)$ of the vector estimate ξ_x by the following equality:

$$\mathrm{D}\xi = \mathrm{E}_{x_0}(q_0^T\Psi(x_0)q_0) + \mathrm{D}_{x_0}(q_0^T\Phi^*(x_0)).$$

Hence the finiteness of the estimate ξ variance follows from the finiteness of the covariance matrix $\Psi(x)$.

We shall consider now the problem of easing the sufficient conditions of unbiasedness and variance finiteness of the vector estimate ξ_x, imposed on the operator $\mathbf{K}$. As it is known, if there is n_0 such that $\| \mathbf{K}^{n_0} \| \leq q < 1$, then the solutions of equations (3.62) and (3.63) can be expressed as the Neumann series:

$$\Phi = \sum_{n=0}^{\infty} \mathbf{K}^n F \quad \text{and} \quad \Phi^* = \sum_{n=0}^{\infty} \mathbf{K}^{*n} H$$

accordingly. It should be noted that the existence of such n_0 is equivalent to the requirement $\rho(\mathbf{K}) \leq q < 1$.

From a physical sense of the solution of equation (3.62) it is clear that the operator $\mathbf{K}$ makes invariant the cone of the Stokes vector functions

$$S_t = \{\Phi(x) = (I_1, I_2, I_3, I_4)^T | I_1 \geq 0, I_2^2 + I_3^2 + I_4^2 \leq I_1^2\}.$$

Therefore we suppose that the phase matrix $P(\omega', \omega, \mathbf{r})$ has properties ensuring the invariances of the operators $\mathbf{K}$ and $\mathbf{K}^*$ into the cone S_t.

Let us consider the problem of unbiasedness of the estimate ξ_x, representable as

$$\xi_x = \sum_{n=0}^{\infty} \Delta_n Q_n H(x_n),$$

where Δ_n is equal to 1 before the first termination of the Markov chain and is equal to 0 after it. It is clear that for the possibility to average the obtained series, the condition $\mathrm{E}\xi_1 < \infty$ is sufficient, where ξ_1 is the vector estimate for the equation of type (3.63) with modules of functional terms, since ξ_1 is a component-wise majorant for the estimate ξ_x. This condition, in turn, is equivalent to the above mentioned requirement $\rho(\mathbf{K}_1) \leq q < 1$. However, proceeding from the condition $H \in S_t$, we may note that in this case the vector function ξ_x is Stokesian and the series for its first component is non-negative. Thus, the inequality

$$|(\sum_{n=0}^{m} \Delta_n Q_n H(x_n))_i| \leq (\sum_{n=0}^{m} \Delta_n H(x_n))_1, \quad i = 2, \dots, 4,$$

holds, which allows us to use the first component of the estimate ξ_x as a majorant for other components. The possibility to average the series for the first component is ensured by the condition $\rho(\mathbf{K}) \leq q < 1$.

In a case $H \notin S_t$ we shall take advantage of the following reasonings.

Let us consider the representation of $H(x)$ as a linear combination of the Stokes vector functions:

$$H(x) = H^{(1)}(x) - H^{(2)}(x),$$

where

$$H^{(1)}(x) = (W, h_2(x), h_3(x), h_4(x))^T, \; H^{(2)}(x) = (W - h_1(x), 0, 0, 0)^T, \; W = 2 \parallel H \parallel_{L_\infty(x)} .$$

Then the estimate ξ_x can be presented as a difference of estimates possessing a property of the Stokes vector functions:

$$\xi_x = \sum_{n=0}^{N} Q_n H(x_n) = \sum_{n=0}^{N} Q_n H^{(1)}(x_n) - \sum_{n=0}^{N} Q_n H^{(2)}(x_n) = \xi_x^{(1)} - \xi_x^{(2)}.$$

From this representation it follows that the series for the first components of the estimates $\xi_x^{(1)}$ and $\xi_x^{(2)}$ are non-negative and have a majorizing property in relation to other components:

$$|(\sum_{n=0}^{m} \Delta_n Q_n H(x_n))_i| \leq |(\sum_{n=0}^{m} \Delta_n Q_n H^{(1)}(x_n))_i| + |(\sum_{n=0}^{m} \Delta_n Q_n H^{(2)}(x_n))_i|$$

$$\leq (\sum_{n=0}^{m} \Delta_n Q_n H^{(1)}(x_n))_1 + (\sum_{n=0}^{m} \Delta_n Q_n H^{(2)}(x_n))_1.$$

From here the validity of the inequalities follows:

$$|\mathrm{E}(\xi_x)_i| \leq |\mathrm{E}(\xi_x^{(1)})_1| + |\mathrm{E}(\xi_x^{(2)})_1| < c < \infty, \ i = 1, \ldots, 4,$$

which allows us to consider proved the following

Statement 1. *Let the operator* $\mathbf{K}^*$ *from equation (3.63) make invariant the cone of Stokes vector functions and* $\rho(\mathbf{K}) \leq q < 1$. *Then for any function* $H(x) \in L_\infty(X)$, *the equality* $\mathrm{E}\xi_x = \Phi^*(x)$ *holds.*

Let us consider the expression:

$$\xi_x \xi_x^T = H(x)\xi_x^T + \xi_x H^T - H(x)H^T(x) + \frac{K^T(x, x_1)}{p(x, x_1)} \xi_{x_1} \xi_{x_1}^T \frac{K(x, x_1)}{p(x, x_1)},$$

which under the condition $\rho(\mathbf{K}_{1,\mathbf{p}}) \leq q_p < 1$ reduced to equation (3.69) (with the use of repeated conditional estimation at fixed x_1 and then over x_1) [49].

If $H(x) \in S_t$, it is easy to see that the series for the component Ψ_{11} of the matrix $\Psi(x) = \mathrm{E}\{\xi_x \xi_x^T\}$ is of constant sign. This allows us to apply to it an operation of varying the summation order and an operation of repeated averaging. In addition there is the inequality $(\xi_x \xi_x^T)_{ij} \leq (\xi_x \xi_x^T)_{11}$, by means of which it is possible to state that the requirement $\rho(\mathbf{K_p}) \leq q_p < 1$ is sufficient for equation (3.69) to be valid.

If $H(x) \notin S_t$, we shall take advantage of the above-stated expansion $\xi_x = \xi_x^{(1)} - \xi_x^{(2)}$, from which by the usual methods the relation

$$|(\xi_x \xi_x^T)_{ij}| \leq (\xi_x^{(1)}[\xi_x^{(1)}]^T)_{11} + (\xi_x^{(1)}[\xi_x^{(2)}]^T)_{11} + (\xi_x^{(2)}[\xi_x^{(1)}]^T)_{11} + (\xi_x^{(2)}[\xi_x^{(2)}]^T)_{11}$$

follows, and $\mathrm{E}(\xi_x^{(i)}[\xi_x^{(j)}]^T)_{11} \leq c_p < \infty$, $i, j = 1, \ldots 4$. This gives us the possibility of formulating the following

Statement 2. *Let the operator* $\mathbf{K}^*$ *from equation (3.63) make invariant the cone of Stokes vector functions and* $\rho(\mathbf{K}) \leq q < 1$, $\rho(\mathbf{K_p}) \leq q_p < 1$. *Then for any function* $H(x) \in L_\infty(X)$, *the covariance matrix* $\Psi(x) = \mathrm{E}\{\xi_x \xi_x^T\}$ *satisfies equation (3.69).*

Thus in a case of non-Stokesian H it is possible to apply the Monte-Carlo method under the condition $\rho(\mathbf{K}) \leq q < 1$. The variances of the studied estimates are finite under the condition $\rho(\mathbf{K_p}) \leq q_p < 1$ (about estimating q_p see below).

Let us consider now the problem of evaluating the derivatives of a solution of integral equation (3.62) with respect to parameters which are included in a scalar factor of the matrix kernel. We shall take advantage of methods of direct derivation of the estimate ξ_x with respect to parameters σ_c and σ_s.

Suppose that scattering and absorption coefficients do not depend on $\mathbf{r}$ and $\sigma = \sigma_s + \sigma_c^{(0)} + \sigma_c$. Suppose the transition density function $p(x', x)$ is independent of σ_c (for instance, it may correspond to any modification of the Monte Carlo method for a medium with $\sigma \equiv \sigma^{(0)} = \sigma_s + \sigma_c^{(0)}$).

After a choice of the recurrent state of the Markov chain we recalculate the weight Q_n according to rule (3.65). By carrying out the appropriate transformations it is easy to obtain that

$$Q_n = \exp\{-\sigma_c L_n\} Q_n^{(0)}$$

where L_n is the full length of the particle trajectory up to the nth collision, $Q_n^{(0)}$ is independent of σ_c. From here

$$\left.\frac{dQ_n}{d\sigma_c}\right|_{\sigma_c=0} = Q'_n = -L_n Q_n,$$

and the derivative of the vector estimate ξ_x looks like:

$$\xi'_x = \sum_{n=0}^{N}(Q'_n H(x_n) + Q_n H'(x_n)) = \sum_{n=0}^{N} Q_n(H'_n - L_n H_n). \tag{3.70}$$

To complete the probability representation of the derivatives $[\varphi_i^*]'_{\sigma_c}$, $i = 1, \ldots, 4$, we use the majorant principle. Since $L_n \le C_0 n$ and

$$|(Q_n H(x_n))_i| \le (Q_n H(x_n))_1, \quad i = 2, 3, 4,$$

$$E|(\Delta_n Q_n H(x_n))_1| \le C_1 q^n, \quad E|(\Delta_n L_n Q_n H(x_n))_1| \le C_2 n q^n,$$

we have

$$|E \sum_{n=0}^{\infty}(\Delta_n Q_n (H'(x_n) - L_n H(x_n)))_i| \le E|\sum_{n=0}^{\infty} \Delta_n Q_n H_n^{(1)}|_1$$

$$+E|\sum_{n=0}^{\infty} \Delta_n Q_n H_n^{(2)}|_1 + E|\sum_{n=0}^{\infty} \Delta_n Q_n L_n H_n|_1 \le C \sum_{n=0}^{\infty} n q^n < \infty.$$

Statement 3. *If the operator* $\mathbf{K}^*$ *from equation (3.63) makes invariant the cone of Stokes vector functions and* $\rho(\mathbf{K}) < 1$, *then the derivative* ξ'_x *of the vector estimate* ξ_x *with respect space to the absorption coefficient* σ_c *is unbiased, i.e.* $\mathrm{E}\xi'_x = \Phi^{*\prime}(x)$.

Let us consider now the task of evaluating the derivative of a solution of equation (3.63) with respect to the scattering coefficient σ_s.

Let $\sigma = \sigma_c + \sigma_s^{(0)} + \sigma_s$ and the transition density function $p(x', x)$ be independent of σ_s (for instance, it may correspond to any modification of the Monte Carlo method for medium with $\sigma \equiv \sigma^{(0)} = \sigma_c + \sigma_s^{(0)}$).

In this case, in view of the above-stated expressions we obtain

$$Q_n = (\sigma_s + \sigma_s^{(0)})^n \exp\{-\sigma_s L_n\} Q_n^{(0)},$$

where $Q_n^{(0)}$ is independent of σ_s. From here

$$\left.\frac{dQ_n}{d\sigma_s}\right|_{\sigma_s=0} = Q'_n = -L_n Q_n + \frac{n}{\sigma_s^{(0)}} Q_n$$

and the expression for the vector derivative estimate ξ'_x with respect to the coefficient σ_s is as follows

$$\xi'_x = \sum_{n=0}^{N}(Q'_n H(x_n) + Q_n H'(x_n)) = \sum_{n=0}^{N} Q_n(H'_n - \hat{L}_n H_n), \tag{3.71}$$

where $\hat{L}_n = L_n - n/\sigma_s^{(0)}$.

Since $L_n \le C_0 n$ and $Q'_n \le n C_1 Q_n$, the unbiasedness proof in this case is similar to the foregoing one.

Statement 4. *If the operator* $\mathbf{K}^*$ *from equation (3.63) makes invariant the cone of Stokes vector functions and* $\rho(\mathbf{K}) < 1$, *then the derivative* ξ'_x *of the vector estimate* ξ_x *with respect to the scattering coefficient* σ_s *is unbiased, i.e.* $\mathrm{E}\xi'_x = \Phi^{*\prime}(x)$.

So, we see that under construction of the algorithms of derivation with respect to the parameters σ_c and σ_s for correctness of the obtained estimates the requirement of medium boundedness and the condition $\rho(\mathbf{K}) \le q < 1$ are sufficient.

Let us consider now the problem of evaluation of the variance of obtained derivation algorithms. For this purpose we shall find an analytical expression for the covariance matrix $\Psi_{\xi'\xi'}(x) = E(\xi'_x[\xi'_x]^T)$.

The following relations are valid:

$$\begin{aligned}\mathrm{E}\xi'_x[\xi'_x]^T &= \mathrm{E}\sum_{n=0}^{N}\sum_{k=0}^{N}(Q'_nH_n + Q_nH')(H_k^T[Q'_k]^T + [H'_k]^TQ_k^T)\\ &= (H'[\Phi^{*\prime}]^T + \Phi^{*\prime}[H']^T - H'[H']^T)(x)\\ &+\mathrm{E}\sum_{n=1}^{N}\sum_{k=1}^{N}(Q'_nH_n + Q_nH'_n)(H_k^T[Q'_k] + [H'_k]^TQ_k^T)\end{aligned}$$

and

$$\begin{aligned}&\sum_{n=1}^{N}\sum_{k=1}^{N}(Q'_nH_n + Q_nH'_n)(H_k^T[Q'_k]^T + [H'_k]^TQ_k^T)\\ &= Q'_1\xi_{x_1}\xi_{x_1}^T[Q'_1]^T + Q'_1\xi_{x_1}[\xi'_{x_1}]^TQ_1^T + Q_1\xi'_{x_1}\xi_{x_1}^T[Q'_1]^T + Q_1\xi'_{x_1}[\xi'_{x_1}]^TQ_1^T.\end{aligned}$$

From here it follows that the covariance matrix $\Psi_{\xi',\xi'}(x) = \mathrm{E}(\xi'_x[\xi'_x]^T)$ is a solution of the equation:

$$\begin{aligned}\Psi_{\xi',\xi'}(x) &= (H'[\Phi^{*\prime}]^T + \Phi^{*\prime}[H']^T - H'[H']^T)(x)\\ &+\int_X \frac{\tilde{K}(x,y)\Psi_{\xi,\xi}(y)\tilde{K}^T(x,y)}{p(x,y)}\mathrm{d}y - \int_X \frac{\tilde{K}(x,y)\Psi_{\xi,\xi'}(y)K(x,y)}{p(x,y)}\mathrm{d}y\\ &-\int_X \frac{K(x,y)\Psi_{\xi',\xi}(y)\tilde{K}(x,y)}{p(x,y)}\mathrm{d}y + \int_X \frac{K^T(x,y)\Psi_{\xi',\xi'}(y)K(x,y)}{p(x,y)}\mathrm{d}y\end{aligned}$$

or in operator form $\Psi_{\xi',\xi'} = \chi + \tilde{\mathbf{K}}_p\Psi_{\xi,\xi} - \check{\mathbf{K}}_p\Psi_{\xi,\xi'} - \hat{\mathbf{K}}_p\Psi_{\xi',\xi} + \mathbf{K}_p\Psi_{\xi',\xi'}$, where

$$[\tilde{\mathbf{K}}\varphi](x) = \int_X |r - r'|K^T(x,x')\varphi(x')\mathrm{d}x' \tag{3.72}$$

for estimating the derivative with respect to the absorption coefficient and

$$[\tilde{\mathbf{K}}\varphi](x) = \int_X \left(|r - r'| - \frac{1}{\sigma_s^{(0)}}\right)K^T(x,x')\varphi(x')\mathrm{d}x' \tag{3.73}$$

for estimating the derivative with respect to the scattering coefficient.

Statement 5. *If the operator* $\mathbf{K}^*$ *from equation (3.63) makes invariant the cone of Stokes vector functions and* $\rho(\mathbf{K}) < 1$, $\rho(\mathbf{K}_p) < 1$, *then for the finiteness of variances of vector estimates (3.70) and (3.71), it is sufficient to require:* $\rho(\tilde{\mathbf{K}}) < 1$, $\rho(\check{\mathbf{K}}_p) < \infty$, $\rho(\tilde{\mathbf{K}}_p) < \infty$, *where the operator* $\tilde{\mathbf{K}}$ *is defined by equalities (3.72) or (3.73) for derivatives with respect to absorption or scattering coefficients respectively.*

Now consider the specified estimation of $\rho(K_p)$ [21]. In comparison with [41] the system of equation (3.63) is adjoint, and therefore

$$K(x,y) = k(x,(\mathbf{r}_y,\omega_y))P^T(\omega,\omega_y)|\mathbf{r}_y - \mathbf{r}|^{-2}\delta(\omega_y - \mathbf{s}),$$

where k is the substochastic density of transition from $\mathbf{r}$ into $\mathbf{r}_y$ along ω_y and

$$P^T(\omega,\omega_y) = L(-i_1)R(\mu)L(\pi - i_2)/2\pi, \quad \mu = (\omega,\omega_y),$$

$$R(\mu) = \begin{pmatrix} r_{11} & r_{12} & 0 & 0 \\ r_{21} & r_{22} & 0 & 0 \\ 0 & 0 & r_{33} & r_{34} \\ 0 & 0 & -r_{43} & r_{44} \end{pmatrix}, \quad L(i) = \begin{pmatrix} 1 & 0 & 0 & 0 \\ 0 & \cos 2i & \sin 2i & 0 \\ 0 & -\sin 2i & \cos 2i & 0 \\ 0 & 0 & 0 & 1 \end{pmatrix},$$

where

$$r_{ij} = r_{ij}(\mu), \quad r_{11} \le 0, \quad \int_{-1}^{+1} r_{11}(\mu)d\mu = 1.$$

It is assumed that the medium is isotropic and P does not depend on $\mathbf{r}$.

To estimate the solution of equation (3.63) by the Monte Carlo method a random variable ξ_x is constructed such that $\mathrm{E}\xi_x = \Phi(x)$. If term-by-term averaging of the series expressing $\xi_x\xi_x^T$ is permissible, then the covariance matrix $\mathrm{E}(\xi_x\xi_x^T) = \psi(x)$ satisfies the matrix-integral equation

$$\psi(x) = \chi(x) + \int \frac{K(x,y)\psi(y)K^T(x,y)}{p(x,y)}dy, \tag{3.74}$$

where $\chi = H\Phi^T + \Phi H^T - HH^T$, and $p(x,y)$ is the density of transition of the Markov chain being modeled, with

$$p(x,y) = p_1(x,(\mathbf{r}_y,\omega_y))p_2(\mu)|\mathbf{r}_y - \mathbf{r}|^{-2}\delta(\omega_y - \mathbf{s})/2\pi,$$

Equation (3.74) is considered in the space $E(R\times\Omega)$ of matrix-valued functions continuous on $R\times\Omega$ with the norm $\| \psi \| = \sup_{i,j,x} |\psi_{ij}(x)|$. Let us denote the matrix-integral operator in equation (3.74) by K_p. If the spectral radius $\rho(K_p) < 1$, then averaging of the matrix $\xi_x\xi_x^T$ mentioned above for $H^T \equiv (1,0,0,0)$ is permissible, as by virtue of properties of the Stokes function [41] we have

$$\xi_{x,1} \ge 0, \quad \mathrm{E}\xi_{x,1}^2 < +\infty, \quad |\xi_{x,i}| \le c\xi_{x,1},$$

and the last inequality is preserved if quantities corresponding to an arbitrary bounded H are substituted in its left-hand side. It is not hard to show that under the condition

$$r_{ij}(\mu)[p_2(\mu)]^{-1/2} \in C[-1,+1], \quad i,j = 1,2,3,4, \tag{3.75}$$

the matrix-integral operator S_p, obtainable from K_p by the substitution $x \to \omega$, $y \to \omega_y$, $p \to p_2$, $K \to P^T$, is completely continuous.

Lemma 3.2.

$$\rho(K_p) \le q\rho(S_p), \quad q = \sup_{\mathbf{r},\omega} \int_0^\infty \frac{k^2(\mathbf{r},(\mathbf{r}+l\omega,\omega))}{p_1(\mathbf{r},(\mathbf{r}+l\omega,\omega))}dl.$$

Proof. Note that in the formulation of the lemma, the integral is taken with respect to the length of the transition from the point $\mathbf{r}$ in the direction ω. In the integral expressing the first component of the nth iteration of the operator K_p we shall successively estimate the integrals with respect to the lengths of the first, second, etc. transitions, with the help of the quantity q. Thus using properties of the Stokes vector we shall obtain the estimate

$$\|K_p^n\|^{1/n} \le c^{-1/n} q \|S_p^n\|^{1/n}.$$

Passage to the limit with respect to n completes the proof. □

An examination of the iteration of the operator K_p shows that the value obtained for the coefficient q is excessive and its real value can coincide with ρL_p, where L_p is the integral operator with the kernel related to $k^2 p_2/p_1$. In every case, if $q < 1/\rho(S_p)$ then the estimates of the form $F^T \xi_x$ have finite variance.

Lemma 3.3. *Let the condition of equation (3.75) be satisfied. Then the operator S_p leaves invariant the cone $T \subset E(\Omega)$ of functions continuous on Ω with values in the class of non-negative definite matrices, and $\rho(S_p) = \lambda_0$, where $\lambda_0 > 0$ is the maximal positive (i.e. conic) eigenvalue of the operator S_p.*

Proof. The first assertion follows from the fact that the transformation $K\psi K^T$ preserves equation the property of non-negative definiteness of a matrix and condition of equation (3.75) guarantees equation continuity of the matrix kernel of system of equation (3.74). Since S_p is completely continuous and the closure of the linear span $\bar{L}(T) = E(\Omega)$, it follows that $\rho(S_p)$ is a positive eigenvalue [13]. On the other hand, complete continuity guarantees the equality $\rho(S_p) = \lambda^*$, where λ^* is the maximal absolute value of the eigenvalues of the operator S_p. Lemma 3.3 is proved. □

We shall now seek the eigenmatrix $\psi^{(0)}(\Omega)$ of the operator S_p in the form of the diagonal matrix with non-negative elements $(1, a_1, a_1, a_2)$ on the diagonal. Direct computations show that the matrix $S_p\psi^{(0)}$ is diagonal; equating its elements with the corresponding elements of the matrix $\lambda_0\psi^{(0)}$, we obtain the system of equations

$$\begin{aligned} c_{11} + c_{12}a_1 &= \lambda_0, \\ c_{12} + (c_{22} + c_{33})a_1 + c_{34}a_2 &= 2\lambda_0 a_1, \\ c_{43}a_1 + c_{44}a_2 &= \lambda_0 a_2, \end{aligned} \tag{3.76}$$

where $c_{ij} = \int [r_{ij}(\mu) r_{ji}(\mu)/p_2(\mu)] \mathrm{d}\mu$.

Lemma 3.4. *If there exists a solution of equation (3.76) with positive components λ_0, a_1 and a_2, then λ_0 is the maximal positive eigenvalue of S_p.*

Proof. Under conditions of the lemma, λ_0 is a positive eigenvalue of S_p, corresponding to the diagonal eigenmatrix ψ_0 with positive diagonal elements. Since ψ_0 is an interior element of the cone T, λ_0 is the maximum positive eigenvalue of K_p [13]. Lemma 3.4 is proved. □

Theorem 3.4. *Under the conditions of Lemmas 3.3 and 3.4, $\rho(S_p) = \lambda_0$, where λ_0 is a positive solution of equation (3.76).* □

The relations we have obtained permit us to solve numerically the problem of minimization of the magnitude of $\rho(S_p)$ as a functional of p_2. In this case it is expedient to consider p_2 from the class

$$p_2(\mu) = c\left[\sum_{i,j=1}^{4} \lambda_{ij}|r_{ij}(\mu)r_{ji}(\mu)|\right], \quad \lambda_{ij} \geq 0,$$

on which is attained the minimax (see [21], Section 4.1) on the coefficients c_{ij} of the operator S_p. The minimax point can be used as the initial point for the search for the optimal point $\{\lambda_{ij}\}$.

The matrix of Rayleigh scattering [41] is a particular case of matrices for which $r_{11} \equiv r_{22}, r_{21} \equiv r_{12}, r_{33} = r_{44}$ and $r_{34} \equiv 0$, where $r_{11}^2 = r_{12}^2 + r_{33}^2$. For such matrices equation (3.76) separates into the system

$$c_{11} + c_{12}a_1 = \lambda_0, \quad c_{12} + (c_{11} + c_{33})a_1 = 2\lambda_0 a_1 \tag{3.77}$$

and the equality $\lambda_1 = c_{33}$, as in the initial system of equation (3.63) the last equation gets detached. If $p \equiv r$, then $c_{11} = 1$, and equation (3.77) gives

$$a_1 = 1/2, \quad \lambda_0 = 1 + c_{12}/2.$$

For Rayleigh scattering

$$r_{11}(\mu) = 3(1+\mu^2)/8, \quad r_{12}(\mu) = 3(1-\mu^2)/8, \quad r_{33}(\mu) = 3\mu/4$$

and for $p_2 \equiv r_{11}$ we obtain

$$\lambda_0 = 1 + (3\pi - 8)/8 = 1.178, \quad \lambda_1 = 0.643.$$

The method of minimization of λ_0 considered in the previous section leads here to minimization of $\lambda_0(a,b)$, the solution of equation (3.77), where

$$p_2(\mu) = c(1 + a\mu^2 + b\mu^4)^{1/2}.$$

In practice, the value $\lambda_0(0.7; 1.0) = 1.171$ turned out to be minimal, and also $1.171 \leq \lambda_0 \leq 1.172$ for all points of the segment $a(t) = 1.7t$, $b(t) = 1.7(1-t)$, $0 \leq t \leq 1$. It is interesting to note that the minimax of the coefficients c is attained here for $a = 2$ and $b = 1$, i.e. for $p_2 \equiv r_{11}$. This explains the high efficiency of the variant $p_2 \equiv r_{11}$.

In conclusion let us note that by iterating K_p with the help of the Monte Carlo method it is possible to estimate the spectral radius approximately. For the case of Rayleigh scattering, such a method of estimation gave a value of $\rho(S_p)$ practically coinciding with λ_0 deduced above. It is convenient to consider as an initial matrix a matrix of the form ll', where $l' = (1, 1/2, 1/2, 0)$. In this case the property $\psi_{11} \leq \psi_{22} + \psi_{33} + \psi_{44}$ is preserved for all iterations, and consequently the resulting eigenmatrix must possess such a property.

By iteration it is also obtained that for a plane layer of optical thickness 2

$$\rho(K_p) = 0.93, \quad q = \rho(K)/\lambda_0 = 0.79, \quad \rho(L_p) = 0.80.$$

These results confirm the assumption stated after Lemma 3.2.

Appendix A
The improvement of random number generators by modulo 1 summation

Numerical statistical modelling becomes even more important when we use multiprocessor systems which allow us to conduct a great number of independent tests. Thus, the requirements for the generators of base random numbers which are uniformly distributed in the interval (0, 1) become more rigid. It is heuristically clear that it is possible to construct physical generators which realize really random numbers. However, it is impossible to ensure the constantly high level of uniformity of the distribution and independence of these numbers.

In this Appendix we substantially refine the familiar method of improving the generators by the so-called congruent summation of random numbers (i.e. modulo 1 summation). We consider the vector version of congruent summation which is especially useful in decreasing the possible dependence of numbers, because it increases the degree of uniformity of multidimensional distributions.

We obtain the power estimates of the deviation of the considered convolutions of multidimensional densities from the density of the uniform distribution in uniform and quadratic metrics. We obtain analogous estimates for discrete distributions which are generally realized in practice when using a finite number of binary digit bits to represent numbers.

From the author's viewpoint the theoretical results obtained strongly suggest that the vector congruent summation allows one to construct precision random number generators. If the computational cost of these generators is high, it is advisable to use them along with standard pseudo-random numbers which are obtained by the multiplicative version of the method of residues.

We consider the theoretical justification of this combination as well as the results of the statistical test of one version of the method of residues, in particular, by using a nonstandard multidimensional test.

A.1. Estimates of the nonuniformity of distributions of the congruent sums of random quantities

Generally speaking, the modulo 1 summation of the independent random quantities

$$Z = \{X_1 + X_2\} \equiv (X_1 + X_2) \pmod 1, \tag{A.1}$$

increases the degree of uniformity of the distribution [22]. In this paper we study this familiar result in some detail.

Let X_1 and X_2 be random quantities with density of the joint distribution $f(x_1, x_2) = p_1(x_1)\, p_2(x_2 \mid x_1)$ on $(0, 1)^2$.

Definition A.1. Summation of the type (A.1) of random quantities is called congruent summation.

The density of the distribution $p(z)$ of the congruent sum Z, $0 \le Z < 1$, is given by

$$p(z) = \int_0^z f(x_1, z - x_1)dx_1 + \int_z^1 f(x_1, 1 + z - x_1)dx_1 .$$

When the random quantities X_1, X_2 are independent, this density is the congruent convolution or the convolution of densities $p_1(x_1)$ and $p_2(x_2)$ on a circle [22]:

$$\begin{aligned} p(z) &= p_1\{*\}p_2(z) \\ &= \int_0^z p_1(x)\, p_2(z - x)\, dx + \int_z^1 p_1(x)\, p_2(1 + z - x)\, dx . \end{aligned} \tag{A.2}$$

If we introduce a new function $s(z, x)$, we can combine the integrals in (A.2):

$$\begin{aligned} p(z) &= \int_0^1 p_1(x)\, s(z, x)\, dx , \\ s(z,\, x) &= \begin{cases} p_2(z - x) , & x < z , \\ p_2(1 + z - x) , & x \ge z . \end{cases} \end{aligned} \tag{A.3}$$

The representation (A.2) admits a vector generalization. We consider independent random vectors $X = (X_1, X_2, ..., X_n)$ and $Y = (Y_1, Y_2, ..., Y_n)$ with the distribution densities

$$p_1^{(n)}(x) = p_1^{(n)}(x_1, x_2, ..., x_n) \text{ and } p_2^{(n)}(y) = p_2^{(n)}(y_1, y_2, ..., y_n) .$$

The coordinate-wise congruent sum of these vectors has the distribution density

$$\begin{aligned} p^{(n)}(z) &= \sum_{\rho_1, ..., \rho_n = 0}^{1} \int_{a_1}^{b_1} \cdots \int_{a_n}^{b_n} p_1^{(n)}(x) \\ &\quad \times p_2^{(n)}(\rho_1 + z_1 - x_1, ..., \rho_n + z_n - x_n)dx_1 ... dx_n , \end{aligned} \tag{A.4}$$

where the intervals of integration are

$$(a_i,\, b_i) = (a_i(\rho_i),\, b_i(\rho_i)) = \begin{cases} (0,\, z_i) , & \rho_i = 0 , \\ [z_i,\, 1) , & \rho_i = 1 , \end{cases} \quad i = 1, ..., n.$$

We denote the hyperparallelepiped $(a_1,\, b_1; a_2,\, b_2; ...; a_n,\, b_n)$ by $S(\rho)$, where $\rho = (\rho_1, ..., \rho_n)$, $\rho_i \in \{0, 1\}$, $i = 1, ..., n$. It is obvious that the expression for $p^{(n)}(z)$ is realated to the following assertion.

Lemma A.1. **(a)** *The following relations hold:*

$$S(\rho) \bigcap S(\rho') = \emptyset, \qquad \rho \ne \rho'$$

$$\bigcup_{\rho_1,\ldots,\rho_n=0}^{1} S(\rho) = (0,\,1)^n\,;$$

(b) *$\forall\, z\, \in\, (0,\,1)^n$ a change of the variables*

$$x' \;=\; \rho + z - x\,, \qquad x\, \in\, S(\rho)\,, \tag{A.5}$$

which is carried for all the admissible values ρ, establishes an automorphism of the hypercube $(0,\,1)^n$.

Proof. The assertions of Lemma A.1 are immediately checked for $n = 1$ and are automatically extended to the multidimensional case, since the hyperparallelepiped $S(\rho)$ corresponds to the different combinations (their number is 2^n) of one-dimensional partitions on different axes. □

By analogy with the one-dimensional case we introduce the function $s^{(n)}(x_1,\ldots,x_n;\, z_1,\ldots,z_n)$ in the hypercube $(0,\,1)^n$:

$$s^{(n)}(x_1,\ldots,x_n;\, z_1,\ldots,z_n) = p_2^{(n)}(\rho_1 + z_1 - x_1,\ldots,\rho_n + z_n - x_n)\,,$$

for $x \in S(\rho)\ \ \forall \rho \;=\; (\rho_1,\ldots,\rho_n)\,,\ \rho_i \in \{0,1\}\,.$

Thus, using Lemma A.1 and relation (A.4), we obtain the formula for the congruent convolution $p^{(n)}(z)$ for the vector version:

$$p^{(n)}(z) \;=\; \int_0^1 \ldots \int_0^1 p_1^{(n)}(x_1,\ldots,x_n)\, s^{(n)}(x_1,\ldots,x_n;\, z_1,\ldots,z_n)\, \mathrm{d}x_1\ldots\mathrm{d}x_n\,.$$

Lemma A.2. *If $p_1^{(n)}(x)$ and $p_2^{(n)}(y)$ are the distribution densities of independent random vector quantities, their congruent convolution admits the representations*

$$\begin{aligned} p^{(n)}(z) \;-\; 1 \;&=\; \int_0^1 \ldots \int_0^1 [p_1^{(n)}(x) - 1]\, s^{(n)}(x;\, z)\, \mathrm{d}x_1\ldots\mathrm{d}x_n \\ &=\; \int_0^1 \ldots \int_0^1 [p_1^{(n)}(x_1,\ldots,x_n) - 1][s^{(n)}(x_1,\ldots,x_n;\, z_1,\ldots,z_n) - 1]\, \mathrm{d}x_1\ldots\mathrm{d}x_n\,, \end{aligned} \tag{A.6}$$

where $x = (x_1,\, x_2,\ldots,x_n)$, $z = (z_1,\, z_2,\ldots,z_n)$.

Proof. By Lemma A.1 and the definition of the function $s^{(n)}$ the integral of $s^{(n)}$ satisfies the equality

$$\int_0^1 \ldots \int_0^1 s^{(n)}(z;\, x)\, \mathrm{d}x_1\ldots\mathrm{d}x_n \;=\; 1\,, \quad z = (z_1,\ldots,z_n) \in (0,1)^n \tag{A.7}$$

at any point z, since $s^{(n)}(z_1,\ldots,z_n;\, x_1,\ldots,x_n)$ becomes $p_2^{(n)}(x_1',\ldots,x_n')$ after the rearrangement of the hyperparallepipeds $S(\rho)$, which is equivalent to the change of the variables (A.5). This proves that relation (A.7) and equalities (A.6) hold, because it is obvious that

$$\int_0^1 [p_1^{(n)}(x) - 1]\, dx_1 ... dx_n = 0.$$

□

In the following we designate:

$$r(p^{(n)}) = \| p^{(n)} - 1 \|_{L^\infty}, \qquad l(p^{(n)}) = \| p^{(n)} - 1 \|_{L^2}.$$

Theorem A.1. *The estimates of the distance between the congruent convolution of the densities of independent random vectors X, Y and the density of the uniform distribution in the metric $L^\infty([0,1]^n)$ have the form:*

$$\begin{aligned} r(p^{(n)}) &\leq r(p_2^{(n)}) \cdot \| p_1^{(n)} - 1 \|_{L^1}, \\ r(p^{(n)}) &\leq l(p_1^{(n)}) \cdot l(p_2^{(n)}), \\ r(p^{(n)}) &\leq \min\{ r(p_1^{(n)}),\, r(p_2^{(n)}) \}. \end{aligned} \tag{A.8}$$

Proof. Using the above notation we obtain the equalities

$$r(s^{(n)}) = r(p_2^{(n)}), \quad l(s^{(n)}) = l(p_2^{(n)}), \quad \forall\, z \in (0,\, 1)^n. \tag{A.9}$$

If we use the second equality from (A.6), relations (A.9), and standard inequalities, we obtain $\forall\, z \in (0,\, 1)^n$:

$$\begin{aligned} | p^{(n)}(z) - 1 | &\leq r(p_2^{(n)}) \cdot \| p_1^{(n)} - 1 \|_{L^1}, \\ | p^{(n)}(z) - 1 | &\leq l(p_1^{(n)}) \cdot l(p_2^{(n)}). \end{aligned} \tag{A.10}$$

It is obvious that the first two inequalities in (A.8) follow from (A.10).

Using Lemma A.1 relations (A.9), and the first equality from (A.6), we get $\forall\, z \in (0,\, 1)^n$ the inequality

$$| p^{(n)}(z) - 1 | \leq \| p_1^{(n)} - 1 \|_{L^\infty} \cdot \int_0^1 \dots \int_0^1 p_2^{(n)}(y_1, \dots, y_n)\, dy_1 \dots dy_n,$$

and this inequality entails the third inequality from (A.8). □

Corollary A.1. *If in the coordinate-wise congruent sum of two random vectors one of the densities is independent of the variable x_i, $1 \leq i \leq n$, the convolution function is independent of the same variable.*

Proof. If we fix the variables $x_1, \dots, x_{i-1}, x_{i+1}, \dots, x_n$ at an arbitrary point $x_1^{(0)}, \dots, x_{i-1}^{(0)}$, $x_{i+1}^{(0)}, \dots, x_n^{(0)}$, we have for the corresponding conditional densities:

$$p_1\{*\}p_2\,(z_i \,|\, x_1^{(0)}, \dots, x_{i-1}^{(0)}, x_{i+1}^{(0)}, \dots, x_n^{(0)}) \equiv 1, \quad 0 \leq x_i \leq 1$$

which can be obtained from relations (A.8) (see also the monograph [22]). □

Thus, the congruent convolution of two vectors, even though $||p_1-1|| \neq 0$, $||p_2-1|| \neq 0$, may lead to $||p-1|| = 0$ if the above uniformity with respect to each variable exists at least for one of the vectors.

Note that in [22] there is an assertion about the weak convergence of congruent convolutions to a constant for $n=1$ (without the estimate of the convergence rate).

The obtained estimate for the quantity $r(p^{(n)})$ cannot be improved, which follows from the example (for $n=1$) of convolution of the 'step' densities $(0 < |c_1|, |c_2| \le 1)$:

$$\begin{aligned} p_1(x) &= 1 + c_1, \; x \in (0, 1/2); \quad p_1(x) = 1 - c_1, \; x \in [1/2, 1), \\ p_2(y) &= 1 + c_2, \; y \in (0, 1/2); \quad p_2(y) = 1 - c_2, \; y \in [1/2, 1). \end{aligned} \tag{A.11}$$

It is easy to verify that in this case the first inequality in (A.8) turns out to be an exact one, since $||p-1||_{L^\infty} = |c_1 c_2|$ holds here. The third relation (A.8) cannot be improved either, which follows from the example of convolution of the two indentical step densities

$$p_1(x) = 1 + c, \qquad x \in \left(0, \frac{1}{1+c}\right)$$

$$p_1(x) = 0, \qquad x \in \left[\frac{1}{1+c}, 1\right)$$

where $c > 1$. In this case

$$||p-1||_{L^\infty} = ||p_1 - 1||_{L^\infty} = c.$$

Taking into account that

$$r(p^{(n)}) = ||\, p^{(n)} - 1 \,||_{L^\infty} = \mathit{vrai} \sup_{x \in (0,1)^n} |\, p^{(n)}(x_1, \ldots, x_n) - 1 \,|,$$

it is sufficient to consider the density that is 'stopped' with respect to one of the variables and independent of the others in order to generalize these examples to a vector version, using the corollary of Theorem A.1.

Note that the relation

$$l(p^{(n)}) \le l(p_1^{(n)}) \cdot l(p_2^{(n)})$$

follows from the second inequality in (A.8).

We now consider the densities $p_1(x)$ and $p_2(y)$, square integrable in the one-dimensional version. They are represented on $[0,1)$ by the Fourier series

$$\begin{aligned} \varphi_1(x;\, a_k, b_k) &= 1 + \sum_{k=1}^{\infty} [a_k \cos(2\pi k x) + b_k \sin(2\pi k x)]\,, \\ \varphi_2(y;\, a'_k, b'_k) &= 1 + \sum_{k=1}^{\infty} [a'_k \cos(2\pi k y) + b'_k \sin(2\pi k y)]\,. \end{aligned} \tag{A.12}$$

Theorem A.2. *If the distribution densities of independent random vector quantities,* $p_1(x)$ *and* $p_2(y)$, *are square integrable, we have the unimprovable equality*

$$l(p) = ||\, p - 1 \,||_{L^2} \le \frac{1}{\sqrt{2}}\, l(p_1)\, l(p_2) \tag{A.13}$$

and the inequality

$$l(p) \; < \; \min\{\, l(p_1)\,,\; l(p_2)\,\}$$

for $l(p_1),\, l(p_2) > 0$.

Proof. If we substitute the series φ_1 and φ_2 for p_1 and p_2, respectively, in the expression of the congruent convolution (A.2), we obtain the expression

$$p(z) \;=\; \int_0^z \varphi_1(x)\,\varphi_2(z-x)\,dx \;+\; \int_z^1 \varphi_1(x)\,\varphi_2(1+z-x)\,dx\,. \qquad \text{(A.14)}$$

Since $\varphi_2(z-x) \equiv \varphi_2(1+z-x)$ for $0 \le 1+z-x < 1$, we can combine the integrals in (A.14), and after the substitution of the series (A.12) we obtain

$$p(z) \;=\; 1 \;+\; \int_0^1 \left[\sum_{k=1}^{\infty} a_k \cos(2\pi k x) + b_k \sin(2\pi k x)\right]$$
$$\times \left[\sum_{k=1}^{\infty} a'_k \cos 2\pi k(z-x) + b'_k \sin 2\pi k(z-x)\right] dx\,.$$

The elementary transformation of the last series leads to the expression

$$\begin{aligned} p(z) \;=\; 1 \;+\; \int_0^1 &\left[\sum_{k=1}^{\infty} a_k \cos(2\pi k x) + b_k \sin(2\pi k x)\right] \\ \times &\left[\sum_{k=1}^{\infty} A'_k(z) \cos(2\pi k x) + B'_k(z) \sin(2\pi k x)\right] dx \end{aligned} \qquad \text{(A.15)}$$

where

$$\begin{aligned} A'_k(z) \;&=\; a'_k \cos(2\pi k z) \;+\; b'_k \sin(2\pi k z)\,, \\ B'_k(z) \;&=\; a'_k \sin(2\pi k z) \;-\; b'_k \cos(2\pi k z)\,. \end{aligned}$$

Since the Fourier series with coefficients $A'_k(z), B'_k(z)$ is also a function which is square integrable for any $z \in [0,1]$, we can apply the generalized Parseval identity [64] to the integral from (A.15), i.e.

$$p(z) \;=\; 1 \;+\; \frac{1}{2} \sum_{k=1}^{\infty} \left[\, A'_k(z)\, a_k + B'_k(z)\, b_k \,\right]. \qquad \text{(A.16)}$$

If we substitute the values of $A'_k(z), B'_k(z)$ in (A.16) and combine the coefficients in front of $\cos(2\pi k z), \sin(2\pi k z)$, we obtain the expression for the congruent convolution

$$p(z) = 1 + \frac{1}{2} \sum_{k=1}^{\infty} (a_k a'_k - b_k b'_k) \cos(2\pi k z) + (a_k b'_k + b_k a'_k) \sin(2\pi k z)\,. \qquad \text{(A.17)}$$

The Parseval identity for (A.17) yields

$$\begin{aligned} \int_0^1 |\, p(z) - 1 \,|^2\, dz \;&=\; \frac{1}{8} \sum_{k=1}^{\infty} [(a_k a'_k - b_k b'_k)^2 + (a_k b'_k + b_k a'_k)^2] \\ &=\; \frac{1}{8} \sum_{k=1}^{\infty} [(a_k a'_k)^2 + (b_k b'_k)^2 + (a_k b'_k)^2 + (b_k a'_k)^2]\,. \end{aligned} \qquad \text{(A.18)}$$

On the other hand,

$$\begin{aligned} \|p_1 - 1\|_{L^2}^2 &= \frac{1}{2}\sum_{k=1}^{\infty}(a_k^2 + b_k^2), \\ \|p_2 - 1\|_{L^2}^2 &= \frac{1}{2}\sum_{k=1}^{\infty}[(a_k')^2 + (b_k')^2]. \end{aligned} \tag{A.19}$$

Therefore, applying the Cauchy-Schwarz-Buniakowskii inequality to (A.18), we obtain the inequality equivalent to (A.13):

$$\|p - 1\|_{L^2}^2 \leq \frac{1}{2}\|p_1 - 1\|_{L^2}^2\|p_2 - 1\|_{L^2}^2. \tag{A.20}$$

Since the identity

$$(a_k a_k')^2 + (b_k b_k')^2 + (a_k b_k')^2 + (b_k a_k')^2 = (a_k^2 + b_k^2)\,[\,(a_k')^2 + (b_k')^2\,] \tag{A.21}$$

holds, the equality in (A.20) can be obtained when all $a_k = a_k' = b_k = b_k' = 0$ except for a single value $k = k_0$, which follows from a simple substitution in (A.18), (A.19). At the same time the presence of at least two harmonics in the expansion of one of the densities does ensure the strict inequality (A.20), which follows from the Cauchy-Schwarz-Buniakowskii inequality.

Let us prove the second inequality. We note that

$$a_k = 2\int_0^1 p_1(x)\cos(2\pi kx)\,dx\,, \quad b_k = 2\int_0^1 p_1(x)\sin(2\pi kx)\,dx$$

or

$$a_k + ib_k = 2\int_0^1 p_1(x)e^{2\pi kix}\,dx$$

and hence,

$$\sqrt{a_k^2 + b_k^2} \leq 2.$$

We assume that for some $k \geq 1$ we have

$$\sqrt{a_k^2 + b_k^2} = 2.$$

In this case $a_k = 2\cos(2\pi kt)$, $b_k = 2\sin(2\pi kt)$, $0 \leq t \leq 1$ and

$$a_k\cos(2\pi kx) + b_k\cos(2\pi kx) = 2\cos[2\pi k(t - x)] = 2\int_0^1 p_1(z)\cos[2\pi k(z - x)]\,dz\,.$$

But

$$\Big|\int_0^1 p_1(z)\,\cos[2\pi k(z - x)]\,dz\Big| < 1 \quad \forall x \in [0, 1]$$

for the measurable function $p_1(z)$, which contradicts the previous equality (for $x = t$) and hence the above assumption. Taking into account that $a_k^2 + b_k^2 < 4$ for all $k \geq 1$, from (A.18), (A.21) we obtain

$$\int_0^1 |p(z)-1|^2 \mathrm{d}z \; < \; \frac{1}{2}\sum_{k=1}^{\infty}[(a'_k)^2+(b'_k)^2] \; = \; l^2(p_2)\,.$$

This relation also holds for the density p_1. This completes the proof of the second inequality in the theorem. □

The density $p(z)$ as a congruent convolution has some peculiarities. Thus, if the initial densities $p_1(x)$ and $p_2(y)$ (in the one-dimensional version) are square integrable under the conditions of Theorem A.2, the function $p(z)$ (A.15) is continuous both on $(0,1)$ and on the entire axis [i.e. $p(0)=p(1)$, and the extension of $p(z)$ is periodic with period 1]. The continuity of the congruent convolution (A.17) follows from the absolute convergence of the series of its Fourier coefficients. Indeed the finiteness of the sums (A.19) entails the finiteness of the sums

$$\sum_{k=1}^{\infty}|a_k a'_k - b_k b'_k| \quad \text{and} \quad \sum_{k=1}^{\infty}|a_k b'_k + b_k a'_k|,$$

since $2\,|a_k a'_k| \le (a_k)^2+(a'_k)^2$.

The constancy condition for one of the densities is a sufficient but not necessary condition for the congruent convolution to be constant.

Lemma A.3. *The congruent convolution for the square integrable densities is the density of the uniform distribution if and only if*

$$S_1 \; = \; \sum_{i=1}^{\infty}(\,|a_i| \; + \; |b_i|\,)\,(\,|a'_i| \; + \; |b'_i|\,) \; = \; 0\,. \qquad \text{(A.22)}$$

Proof. The convergence of the series (A.22) which is formed from the coefficients of their expansions into the Fourier series (A.12) follows from the conditions for the square integrable densities $p_1(x)$ and $p_2(y)$.

The direct analysis of equality (A.17) shows that $p(z)\equiv 1$ follows from $S_1=0$. Let the coefficients a_k, a'_k, b_k, b'_k satisfy the conditions

$$a_k a'_k - b_k b'_k \; = \; 0\,, \qquad a_k b'_k + b_k a'_k \; = \; 0\,. \qquad \text{(A.23)}$$

We show that in this case the coefficients satisfy the equivalent equalities

$$(|a_k|+|b_k|)\,(|a'_k|+|b'_k|)=0\,, \quad (a_k^2+b_k^2)\,[\,(a'_k)^2+(b'_k)^2\,]=0\,. \qquad \text{(A.24)}$$

Indeed, let one of the four coefficients a_k, a'_k, b_k, b'_k, for example, a_k be zero. Then (A.23) leads to $b_k\,a'_k=0$, $b_k\,b'_k=0$, and either $b_k=0$ or $a'_k=0$, $b'_k=0$; hence we obtain (A.24). If all the four coefficients are different from zero, we obtain from (A.23), for example, $a'_k=b_k\,b'_k/a_k$ and $(a_k^2+b_k^2)\,b'_k=0$, which contradicts the above assumption. Thus, $S_1\equiv 0$ follows from $p(z)\equiv 1$. □

It is not difficult to show that the congruent convolution of the density $P_1(c)$ from (A.11) for $c\neq 0$ and the density

$$24y(1/2-y),\quad y\in(0,1/2);\quad 24(y-1/2)(1-y),\quad y\in[1/2,1),$$

is the density of the uniform distribution on (0, 1), which shows that Lemma A.3 is non-trivial.

As examples of the application of the method of modulo 1 summation of random quantities we give the expressions for the congruent convolutions of three typical distributions.

(1) We consider the density $p(x,\sigma)$ of a normal distribution with zero expectation and variance σ. We find its mapping onto the unit interval:

$$\begin{aligned} p^{(1)}(y,\sigma) &= \frac{1}{\sigma\sqrt{2\pi}}\sum_{m=-\infty}^{\infty}\exp\left[-\frac{(y+m)^2}{2\sigma^2}\right] \\ &= 1\,+\,2\sum_{k=1}^{\infty}\cos(2\pi ky)\,\exp[-2\pi^2k^2\sigma^2],\quad 0\le y\le 1. \end{aligned} \tag{A.25}$$

This expression (see [64]) is a continuous function on the entire axis. The analogous mapping for the normal density $p(x,a,\sigma)$ with nonzero expectation a is

$$p^{(1)}(y,a,\sigma)=1+2\sum_{k=1}^{\infty}\cos[2\pi k(y-b)]\,\exp\,[(-2\pi^2k^2\sigma^2],\quad b=\{a\}. \tag{A.26}$$

The congruent convolution of the two densities of normal distributions with parameters $(a,\,\sigma_1)$ and $(b,\,\sigma_2)$ is determined by

$$p_1\{*\}p_2(z)\;=\;p^{(1)}(z,c,\sqrt{\sigma_1^2+\sigma_2^2}),\quad c=\{a+b\}. \tag{A.27}$$

The functions (A.25) - (A.27) are continuous. From (A.25), (A.26) it follows that the considered mappings even for rather small σ may be close to the uniform distribution in the given metrics. Relation (A.27) shows that as a result of the congruent summation this closeness increases, since the degree of uniformity increases as the variance increases.

(2) For the density $Ex(x,\lambda)=\lambda\exp(-\lambda x)$, $x\ge 0$, of the characteristic distribution, the mapping onto the unit interval is determined by

$$Ex^{(1)}(y,\lambda)\;=\;\frac{\lambda e^{\lambda}}{e^{\lambda}-1}\,\exp(-\lambda y),\quad \lambda>0,\quad 0\le y\le 1.$$

The congruent convolution of two such densities with parameters λ_1, λ_2 for $\lambda_1\ne\lambda_2$ is

$$\begin{aligned} p(z,\lambda_1,\lambda_2) &= Ex_1\{*\}Ex_2(z) \\ &= \frac{\lambda_1\lambda_2e^{\lambda_2+\lambda_1}}{(\lambda_1-\lambda_2)(e^{\lambda_2}-1)(e^{\lambda_1}-1)}[e^{-\lambda_2 z}-e^{-\lambda_1 z}+e^{-\lambda_2-\lambda_1 z}-e^{-\lambda_1-\lambda_2 z}]. \end{aligned}$$

Note that this function, unlike $Ex^{(1)}(y,\lambda)$, may be continuously extended to the entire axis, since $p(0,\lambda_1,\lambda_2)\;=\;p(1,\lambda_1,\lambda_2)$. We have a different expression for the congruent convolution for $\lambda_1=\lambda_2$.

(3) In a number of widely used deterministic algorithms for simulating random numbers, the points are located on a family of parallel straight lines lying on the plane [19]. This justifies the consideration of density as a saw-tooth function

$$p_\alpha(x,n) = 1+2\alpha\{nx\}-\alpha, \quad 0<\alpha\le 1, \quad 0\le x\le 1.$$

The values of deviation of the densities $p_\alpha(x,n)$, $p_\beta(y,m)$ and their congruent convolution from the density of the uniform distribution in the metrics L^2 and L^∞ are related respectively by the equations

$$l=(n,m)^2\frac{l_\alpha\, l_\beta}{\sqrt{5}\, n\, m}, \qquad r=(n,m)^2\frac{r_\alpha\, r_\beta}{3\, n\, m}, \qquad n,m\in\mathbb{N},$$

where (n,m) is the greatest common divisor of numbers n and m.

We now consider the repeated congruent summation.

Definition A.2. The distribution density of the vector quantity

$$Y = (\{X_1^{(1)}+X_1^{(2)}+\ldots+X_1^{(m+1)}\},\ldots,\{X_n^{(1)}+X_n^{(2)}+\ldots+X_n^{(m+1)}\})$$

is called the congruent m-fold convolution of densities of vector random quantities $X^{(1)},\ldots,X^{(m+1)}$.

According to the definitions, the multiple operations of the congruent summation are commutative and associative. It is logical that multiple congruent convolutions also have these properties. Therefore, when investigating them, it is sufficient to consider sequential convolutions with initial densities. Thus, if the distribution density of the n-dimensional vector $p_{(m)}^{(n)}(\cdot)$ is the m-fold convolution, then

$$p_{(m)}^{(n)}(\cdot) = p_{(m-1)}^{(n)}\{*\}p_{m+1}^{(n)}(\cdot) \equiv p_{(m-k)}^{(n)}\{*\}p_{(k-1)}^{(n)'}(\cdot), \quad 1\le k\le m,$$

where $p_i^{(n)}(\cdot)$ are the initial densities, $1\le i\le m+1$, $p_{(0)}^{(n)}(\cdot)\equiv p_1^{(n)}(\cdot)$, and $p_{(m-k)}^{(n)}(\cdot)$, $p_{(k-1)}^{(n)'}(\cdot)$ are the congruent convolutions of $m-k+1$ and k first and final initial densities, respectively.

Theorem A.3. *If for $i=1,2,\ldots,m+1$, $m\ge 1$, the distribution densities $p_i(\cdot)$ of independent scalar random quantities are square integrable and $p_{(m)}(\cdot)$ is their congruent m-fold convolution, the unimprovable inequality*

$$l(p_{(m)}) \le 2^{-m/2}\prod_{i=1}^{m+1} l(p_i), \qquad l(p_i)>0 \tag{A.28}$$

holds, as well as the inequality

$$r(p_{(m)}) \le 2^{-(m-1)/2}\prod_{i=1}^{m+1} l(p_i).$$

Proof. It is clear that we must prove only that the first inequality is unimprovable. By Theorem A.2 the equality in (A.28) for $m=1$ is obtained only if in the expansions (A.12) all the coefficients except for the coefficient with the subscript k_0 are zero. From (A.17) and Lemma A.3 it follows that the congruent convolution has the same harmonic defined by k_0. Therefore we admit the recurrent transition $m\to m+1$, which results in the equality in (A.28).

The second inequality in Theorem A.3 follows from relation (A.28) and the second inequality in (A.28). □

A.2. Congruent sums of grid random quantities

In order to carry out Monte Carlo calculations we approximately realize a random number from (0, 1), given the finite number of binary digit bits.

Therefore we consider the version of grid (for a grid of volume N) discrete random quantities X, Y with possible values $\{i/N\}$, where i are natural numbers from the set $\{1, ..., N\}$. We represent the corresponding probabilities as

$$\begin{aligned} p_1(i) &= P\{X = i/N\} = \frac{1+c_i}{N}, \\ p_2(j) &= P\{Y = j/N\} = \frac{1+d_j}{N}, \end{aligned} \tag{A.29}$$

where

$$-1 < c_i,\, d_i < N-1, \quad \sum_{i=1}^{N} c_i = 0, \quad \sum_{j=1}^{M} d_j = 0.$$

The values

$$l(p_1) = \left\{\frac{1}{N}\sum_{i=1}^{N} c_i^2\right\}^{1/2}, \quad l_1(p_1) = \frac{1}{N}\sum_{i=1}^{N} |c_i|, \quad r(p_1) = \max_{1\le i\le N} |c_i| \tag{A.30}$$

are the characteristics of the deviation of the distribution of the quantity X from the discrete uniform distribution in metrics

$$l^2(\{1,...,N\}), \quad l^1(\{1,...,N\}), \quad \text{and} \quad c(\{1,...,N\}),$$

respectively.

Let us estimate the deviations of the distribution of the congruent sum $Z = \{X+Y\}$ from the uniform one.

Theorem A.4. *If $\{p_1(i)\}$ and $\{p_2(j)\}$ are the distributions of independent grid random quantities and $\{p(m)\}$ is their congruent convolution, then*

$$r(p) \le r(p_2) l_1(p_1), \quad r(p) \le l(p_1) l(p_2), \quad r(p) \le \min\{r(p_1), r(p_2)\}. \tag{A.31}$$

Proof. The discrete analogue of the congruent convolution (A.2) may be represented as

$$p(m) = \sum_{k=1}^{m-1} p_2(m-k)\, p_1(k) + + \sum_{k=m}^{N} p_2(N+m-k)\, p_1(k), \quad m = 1, ..., N. \tag{A.32}$$

As was done in (A.3), we introduce the function $s(i, m)$ and obtain the relations

$$p(m) = \sum_{i=1}^{N} p_1(i)\, s(i, m), \quad s(i, m) = \begin{cases} p_2(m-i), & i = 1, 2, ..., m-1 \\ p_2(N+m-i), & i = m, m+1, ..., N \end{cases} \tag{A.33}$$

$$N p(m) = 1 + \sum_{i=1}^{N} [N p_1(i) - 1]\, s(i, m) = 1 + \frac{1}{N}\sum_{i=1}^{N} [N p_1(i) - 1]\, [N s(i, m) - 1].$$

Using the standard inequalities, we obtain relations (A.31) from equalities (A.33) in the same way as in the continuous case. We can show that inequalities (A.31) are unimprovable, using, as an example, the case of even N in which the conditions for the probabilities (A.29) for $1 \le i \le N/2$ are satisfied:

$$c_{2i-1} = c\,,\ c_{2i} = -c\,,\ d_{2i-1} = d\,,\ d_{2i} = -d\,,\quad c,\, d \neq 0\,.$$

It is easy to see that after we substitute these values of the probabilities in (A.32), we obtain the values $p(m)$ for which relation (A.31) are exact equalities. □

Theorem A.4 admits the generalization to the vector random quantities whose possible values have the form $(i_1/N,\ i_2/N,\ \dots,\ i_n/N)$ and constitute an n-dimensional grid of volume N^n. In this case the values of the probabilities are

$$p_1^{(n)}(i) = \frac{1+c_i}{N^n},\quad p_2^{(n)}(j) = \frac{1+d_j}{N^n},\quad \sum_{i_k=1}^{N} c_{i_1\dots i_n} = \sum_{j_k=1}^{N} d_{j_1\dots j_n} = 0,$$

where $k = 1, \dots, n$ and $i = (i_1, \dots, i_n)$, $j = (j_1, \dots, j_n)$ are vectors.

By analogy with (A.33) and the vector version for continuous distributions we introduce the function $s^{(n)}(i;\, m)$ and obtain the expressions for the congruent convolution of the vector distributions:

$$\begin{aligned} p^{(n)}(m) &= \sum_{i_1,\dots,i_n=1}^{N} p_1^{(n)}(i)\, s^{(n)}(i;\, m) \\ N^n p^{(n)}(m) - 1 &= N^{-n} \sum_{i_1,\dots,i_n=1}^{N} [N^n p_1^{(n)}(i) - 1][N^n s^{(n)}(i;\, m) - 1] \end{aligned} \tag{A.34}$$

where $i = (i_1, \dots, i_n)$ and $m = (m_1, \dots, m_n)$.

By analogy with the one-dimensional case considered, the values

$$l(p^{(n)}) = \left\{ \frac{1}{N^n} \sum_{i_1,\dots,i_n=1}^{N} c_{i_1,\dots,i_n}^2 \right\}^{1/2},$$

$$l_1(p_1^{(n)}) = \frac{1}{N^n} \sum_{i_1,\dots,i_n=1}^{N} |c_{i_1,\dots,i_n}|,\quad r(p_1^{(n)}) = \max_{1\le i_1,\dots,i_n\le N} |c_{i_1,\dots,i_n}|$$

are the characteristics of the deviation of the distribution of an n-dimensional vector from the discrete uniform distribution in the metrics

$$l^2(\{1, \dots, N\})^n,\quad l^1(\{1, \dots, N\})^n,\quad \text{and}\quad c(\{1, \dots, N\})^n,$$

respectively. Using the above expressions, from equality (A.34) we obtain the estimates of the form (A.31) for the independent grid random vectors:

$$r(p^{(n)}) \le r(p_2^{(n)}) \cdot l_1(p_1^{(n)}),\qquad r(p^{(n)}) \le l(p_1^{(n)}) \cdot l(p_2^{(n)}),$$
$$r(p^{(n)}) \le \min\{ r(p_1^{(n)}),\, r(p_2^{(n)}) \}.$$

That these inequalities are unimprovable follows from the examples for the one-dimensional case and the additional arguments for $n > 1$ by analogy with the version of continuous distributions.

Of practical importance also is the version in which one of the congruently summed quantities is a discrete grid quantity of the form $a_i = (i-1)/N$ with probabilities $p_1(i) = (1 + c_i)/N$, and the other has the density $p_2(x)$, $x \in (0, 1)$.

As in (A.2), (A.33), we again introduce the function $s(x,\, z)$ for $n = 1$, which may be conveniently represented here as $s(a_i,\, z)$; then the congruent convolution has the form:

$$p(z) = \frac{1}{N}\sum_{i=1}^{N}(1+c_i)\, s(a_i,\, z), \quad s(a_i,\, z) = \begin{cases} p_2(z - a_i), & a_i < z\,, \\ p_2(1 + z - a_i), & a_i \geq z. \end{cases}$$

It is clear that relations of the form (A.8), generally speaking, cannot hold here, because an equality of the type (A.33) does not hold in this case, since

$$\frac{1}{N}\sum_{i=1}^{N} s(a_i,\, z) \not\equiv 1\,.$$

However, the convolution of a continuous distribution with a discrete one provides an approximation to the uniform distribution.

Proposition A.1. *Under the conditions considered, the following relation holds:*

$$r(p) \leq r(p_2)\,.$$

Proof. The equality

$$p(z) = 1 + \frac{1}{N}\sum_{i=1}^{N}(1+c_i)\,[\,s(a_i,\, z) - 1\,]\,,$$

holds, and $|\, s(a_i,\, z) - 1\,| \leq r(p_2)\,,\ \forall\, z \in (0,\, 1)^n$. □

This assertion is automatically generalized to the multidimensional case and to the version in which the second convolute distribution is also discrete but is determined on a denser grid, i.e. for example, when a great number of binary digit bits are used to represent the possible values.

In the case under consideration it is advisable to use the modification of congruent summation. We denote by X a grid random quantity (for a grid of volume N) and by Y a continuous value (or a grid value for a grid whose volume is larger than N). Let

$$Y_1 = [\mathit{entier}\,(N\,Y)]/N\,, \quad Y_2 = \{N\,Y\}/N\,,$$

and $Y = Y_1 + Y_2$. It is clear that in order to approach the uniform distribution it is better to use the random quantity

$$Z = \{X + Y_1\} + Y_2\,,$$

rather than the congruent sum $\{X + Y\}$, since the characteristics of the distribution of the quantity $\{X + Y_1\}$ are estimated by Theorem A.4.

A.3. Improvement in the random number generators by congruent summation

(1) It is obvious that the above assertions may be applied to analyse the degree of uniformity of the distribution of a deterministic sequence of numbers from (0, 1) which are obtained by the congruent summation of weakly connected initial sequences; instead of the probabilities we are to consider the relative frequencies of different values. This somehow

justifies theoretically the improvement in the quality of pseudorandom numbers in the process of congruent summation [72]. The results of the present work show that particular attention should be given to the independence of the summed sequences (in the sense of the possibility to factor common frequencies [19]).

(2) We consider different possibilities to use congruent summation to improve the random number generators.

The generated random numbers may be repeatedly congruently summed. Theorem A.3 and the one-dimensional version of Theorem A.1 show that the convergence to the uniform distribution is realized at power rate. It is clear that the possible weak dependence of the initial numbers is of little importance and rapidly weakens, which may be rigorously justified for the 'weakly Markov' model of the generator.

The characteristics of the multidimensional uniformity of the numbers produced by the random number generators do not always meet the requirements when using the method of statistical tests in practice. The termwise congruent summation of the random numbers produced by two independent generators seems to be especially efficient.

Let independent random sequences $X_1, X_2, \ldots$ and $Y_1, Y_2, \ldots$ be given for which under the conditions of Theorem A.1 the distribution densities $p_1^{(n)}(x_1, \ldots, x_n)$ and $p_2^{(n)}(y_1, \ldots, y_n)$ of the vectors

$$X^{(n)} = (X_1, X_2, \ldots, X_n) \quad \text{and} \quad Y^{(n)} = (Y_1, Y_2, \ldots, Y_n),$$

respectively, are determined for arbitrary $n \geq 1$.

Theorem A.5. *For the distribution density of the congruent sum of random vectors $X^{(n)}$, $Y^{(n)}$ the following limiting relation holds:*

$$\lim_{n\to\infty} \|1 - p^{(n)}\|_{L^\infty} \leq \lim_{n\to\infty} \|1 - p_1^{(n)}\|_{L^\infty} \cdot \lim_{n\to\infty} \|1 - p_2^{(n)}\|_{L^1}. \tag{A.35}$$

Proof. Since the density $p^{(n)}(x_1, \ldots, x_n)$ is obtained from $p^{(n+1)}(x_1, \ldots, x_n, x_{n+1})$ by averaging over the variable x_{n+1}, the relations

$$\|1 - p^{(n)}\|_{L^\infty} \leq \|1 - p^{(n+1)}\|_{L^\infty}, \quad \|1 - p_2^{(n)}\|_{L^1} \leq \|1 - p_2^{(n+1)}\|_{L^1},$$

hold, and the limits in (A.35) exist. Thus, the first inequality in (A.8) holds for the limiting case as well. □

From relation (A.35) follows a substantial increase in the degree of multidimensional uniformity and thereby in the characteristics of independence of the obtained random numbers provided that the initial distributions are sufficiently satisfactory. If there is no guarantee that this condition will be satisfied, according to Theorems A.3 and A.4 the situation can be improved by preliminary multiple congruent summation. For example, if the sequences $\{X_i\}$ and $\{Y_i\}$ are previously separately congruently summed in pairs, by Theorems A.1 and A.2 we obtain the estimate

$$r(p) \leq \frac{1}{2}\, l^2(p_1) \cdot l^2(p_2)$$

for the one-dimensional distribution of the congruent sum of the new sequences.

(3) The samples from $[0, 1)$ by the uniform distribution are generally modelled in practice by recurrent sequences of pseudorandom numbers $X = \{x_i\}$, for example, by the congruent multiplicative algorithm

$$n_{i+1} \equiv L\, n_i \pmod M\,, \qquad x_i = n_i/M\,; \quad 1 < L,\, M \in \mathbb{N}\,. \tag{A.36}$$

The feasibility of its application, in particular, is based on the properties of uniformity of the distribution of the sequence

$$\varepsilon_n = \{L^n \varepsilon_0\} \equiv L^n \varepsilon_0 \pmod 1 \tag{A.37}$$

for almost all values $\varepsilon_0 \in (0, 1)$ [19], [23]. Besides, Sobol in [70] shows that if ε_0 is a random quantity uniformly distributed in (0, 1), the random vector $(\varepsilon_0, \{L\varepsilon_0\}, ..., \{L^{n-1}\varepsilon_0\})$ as $L \to \infty$, is asymptotically uniformly distributed in $(0, 1)^n$.

We given the important results that justifies the application of sequences of the from (A.37). Let $f(x)$ be a function specified and Riemann integrable in a unit s dimensional hypercube $\mathcal{D}_s$. We set

$$\varphi_L(x) = f(x, \{L_1 x\}, \{L_1 L_2 x\}, ..., \{L_1 ... L_{s-1} x\})\,,$$

where $L_1, ..., L_{s-1}$ are natural numbers greater than unity and $L = \min\{L_1, ..., L_{s-1}\}$. We denote

$$J = \int\limits_{\mathcal{D}_s} f(x) \mathrm{d}x \quad J_L = \int\limits_0^1 \varphi_L(x) \mathrm{d}x\,.$$

Theorem A.6. *[70] If under the assumptions made for some $L_0 \le 0$ at all $L > L_0$ all $\varphi_L(x)$ are Riemann integrable in (0, 1), then $\lim\limits_{L\to\infty} J_L = J$.*

If $f(x)$ satisfies the Lipschitz condition in $\mathcal{D}_s$:

$$|f(x') - f(x'')| \le L_f \sum_{i=1}^{s} |x_i' - x_i''|^\gamma, \quad 0 < \gamma \le 1\,,$$

then for all $L > L_0$ we have

$$\left| \int\limits_{\mathcal{D}_s} f(x) \mathrm{d}x - \int\limits_0^1 \varphi_L(x) \mathrm{d}x \right| \le \frac{(s-1) L_f}{L^\gamma}\,.$$

Note that the version of Theorem A.6 with $L_1 = ... = L_{s-1} = L$ corresponds to the algorithm (A.36).

Thus, there is reason to hope that we can construct a sufficiently economical and satisfactory (in the sense of multidimensional uniformity) random number generator if to model a successive experiment we use in (A.36) the random number ε_0 produced by a reliable physical generator. However, the substantial drawback of the generators is the possible unsatisfactory uniformity of the distribution of the generated numbers. The results of the work show that this drawback can be effectively avoided by congruent summation.

(4) The sequences of numbers that are used in methods of statistical modelling as samples from interval (0, 1) by the uniform distribution need a thorough check. We make general remarks about this problem which is discussed in detail, for example, in [19].

Let $\alpha_1, \alpha_2, ...$ be a sequence of 'real' sample values from (0, 1) by the uniform distribution and the vectors $\eta_1^{(k)}, \eta_2^{(k)}, ..., \eta_n^{(k)}, ...$ are obtained from it as follows:

$$\eta_1^{(k)} = (\alpha_1, ..., \alpha_k), ..., \eta_n^{(k)} = (\alpha_{k(n-1)+1}, ..., \alpha_{nk}), ...$$

These vectors uniformly fill the unit k-dimensional cube with unit probability, i.e. the rate at which a vector appears in any rectangular domain of the cube tends to the volume of this domain as $n \to \infty$. The infinite sequences of numbers which have these properties are called the k-uniform sequences. The following relations hold for them:

$$\frac{1}{N}\sum_{n=1}^{N} f(\eta_n^{(k)}) \longrightarrow \int_0^1 ... \int_0^1 f(x_1, ..., x_k)\, dx_1 ... dx_k,$$

where f is any Riemann integrable function. Therefore, it is very important to check the k-uniformity of pseudorandom numbers. Since we are to model randomness, a check should be done by statistical criteria, for example, by the criterion χ^2. This is done as follows. We divide the k-dimensional cube into $s = r^k$ identical cubes of volume r^{-k}. Let $m_1, ..., m_s$ be the number of points from the sequence $\eta_1^{(k)}, ..., \eta_N^{(k)}$ which fall on the corresponding parts $(m_1 + ... + m_s = N)$. We may ask whether the same or greater deviation from the uniform distribution $m_1 = m_2 = ... = m_s = N s^{-1}$ is possible for a 'true' random sequence.

For sufficiently large N the quantity

$$\bar{\chi}^2_{s-1} = \frac{1}{N s^{-1}} \sum_{i=1}^{s} (m_i - N s^{-1})^2$$

is approximately distributed in the same way as χ^2_{s-1} (chi-square with $s-1$ degrees of freedom) if the vectors $\eta_1^{(k)}, ..., \eta_N^{(k)}$ are assumed to be independent and uniformly distributed. Thus, if $\bar{\chi}^2_{s-1} > \chi^2_{s-1}(p)$, where χ^2_{s-1} is determined by the equation

$$P\{\chi^2_{s-1} \geq \chi^2_{s-1}(p)\} = p \ll 1,$$

it is doubtful whether the above assumption is valid. The quantity $\chi^2_{s-1}(p)$ is called the confidence limit with significance level p and is determined by tables for χ^2.

We can check the uniformity for large k if we consider the special partionings of the unit k-dimensional cube and use, in particular, modelling functions $\xi_i = \varphi_i(\alpha_1, ..., \alpha_n)$. However, we can generally obtain well-founded conclusions from the numerical data only if $\xi_1, \xi_2, ...$ are independent and the above assumption is valid. Thus, we have a problem of constructing the appropriate systems of independent functions $\xi_i(\alpha^{(1)}, ..., \alpha^{(k)})$ with the known distribution.

For this purpose we consider the sequence

$$\beta_i = F^{-1}(\alpha_i), \qquad i = 1, 2, ...$$

rather than $\{\alpha_i\}$, where F is the function of a standard normal distribution; the corresponding k-dimensional vectors are denoted by $\zeta_n^{(k)}$, $(n = 1, 2, ...)$. The hypothesis for the normality of a component of the vector $\zeta^{(k)}$ is equivalent to the hypothesis for the uniformity of the corresponding component of the vector $\eta^{(k)}$.

In order to check the multidimensional normality of the vector $\zeta^{(k)}$ we can use the familiar characteristic property of the multidimensional normal distribution: if any linear combination of the components of the vector $\zeta^{(k)}$ is distributed normally, $\zeta^{(k)}$ is also distributed normally.

Since it is difficult to check the distribution of a large number of linear combinations, it is advisable to consider only the combinations which are independent, for example, the components of the orthogonal transformation $\zeta^{(k)}$. In particular, we can use the orthogonal transformation which is used in statistics: $\zeta^{(k)} = A\zeta^{(k)}$, where

$$A = \left\| \begin{array}{cccccc} \frac{1}{\sqrt{1\cdot 2}} & -\frac{1}{\sqrt{1\cdot 2}} & 0 & \dots & 0 & 0 \\ \frac{1}{\sqrt{2\cdot 3}} & \frac{1}{\sqrt{2\cdot 3}} & -\frac{2}{\sqrt{2\cdot 3}} & \dots & 0 & 0 \\ \dots & \dots & \dots & \dots & \dots & \dots \\ \frac{1}{\sqrt{k(k-1)}} & \frac{1}{\sqrt{k(k-1)}} & \frac{1}{\sqrt{k(k-1)}} & \dots & \frac{1}{\sqrt{k(k-1)}} & \frac{-(k-1)}{\sqrt{k(k-1)}} \\ \frac{1}{\sqrt{k}} & \frac{1}{\sqrt{k}} & \frac{1}{\sqrt{k}} & \dots & \frac{1}{\sqrt{k}} & \frac{1}{\sqrt{k}} \end{array} \right\| .$$

In order to check the distribution of a component of the vector $\xi^{(k)}$ we divide the real axis into r equally probable parts according to the standard normal law. Then, after we analyse all the realizations of $\xi^{(k)}$, we can obtain k values of the Pearson criterion: $\chi^2_{r-1,1}, \dots, \chi^2_{r-1,k}$, which are independent if the above assumption is valid. Consequently, we can use the value of the general criterion

$$\chi^2_s = \chi^2_{r-1,1} + \dots + \chi^2_{r-1,k}, \qquad s = (r-1)k .$$

The corresponding test is called the test of k-normality.

If this test yields a satisfactory result, it is advisable to realize a multidimensional test of independence [3] whose idea is that the random quantity

$$\psi_N^{(k)} = -\left(N - \frac{2k+11}{6}\right) \ln V_N^{(k)}$$

asymptotically (as $N - (2k+11)/6 \to \infty$) has the distribution χ^2_f with the number of degrees of freedom $f = k(k-1)/2$ if the components of the normal vector $\zeta^{(k)}$ are independent ($V_N^{(k)}$ is the determinant of a sample normalized covariance matrix). The quantity $\Psi_N^{(k)}$ may thus be used to check the main hypothesis. This test is called the correlation test.

Under the assumption that $\zeta^{(k)}$ is normal, the quantity $V_N^{(k)}$ is the ratio of similitude and therefore the corresponding criterion is consistent, i.e. the probability to reject the zero hypothesis tends to unity as $N \to \infty$ if this hypothesis is not true. We also know that this criterion is rigorously unbiased relative to the normal alternatives, i.e. the above probability is greater than the significance level. Consequently, the criterion becomes more informative if the multidimensional normality of the vector $\zeta^{(k)}$ is checked, for example, by the above test of k-normality.

We can consider the last two tests as a whole as a nonstandard test of k-uniformity, since the positive result here is a complicated consequence of k-uniformity and is closely related with it.

(5) For the past thirty years, problems which may be considered as an estimate of integrals of high or even countable multiplicity over complicated domains of a unit hypercube have been solved by the Monte Carlo method with the aid of the multiplicative generator (A.36) with multiplier $L = 5^{17}$ and modulus of a congruence $M = 2^{40}$. The results were repeatedly compared with other calculated and experimental data. They were always in reasonably good agreement.

The results of the checking of the initial sequence of volume 10^6 for this generator at $n_0 = 1$ by an extensive system of tests were quite satisfactory [19]. This system included a wave test, a test of series, the above direct test of k-uniformity by the chi-square criterion for $k = 1, 2, 3, 4$ when each edge of the cube was divided into $r = 10$ parts as well as the above nonstandard test of k-uniformity (i.e. the test of k-normality along with the correlation test) for the values of k:

$$1,\ p_0,\ p_1,\ \ldots,\ p_i,\ \ldots,\ p_{29},\ 499,\ 1009\,;$$

here, $\{p_i\}$ are prime numbers in the interval (2, 113).

The results obtained (see [19]) were comprehensively analysed but the significant values of the criteria used were not obtained. Note that the multidimensional tests applied to the unsatisfactory generator (A.36) with parameters $L = 5, M = 2^{40}$ yielded the significant values of the criteria.

To correlate the estimates in different versions when solving some problems it is advisable to use a special order of pseudorandom numbers which is readily realized by a multiplicative algorithm. The sequence $\{\alpha_n\}$ is divided into sufficiently large equal parts that are used for modelling the trajectories of the corresponding numbers. The test of this modification for the part of volume 1024 for the generator (A.36) with $L = 5^{17}$, $M = 2^{40}$ yielded positive results. We used the above system of test and carried out a large numbers of check calculations for complex multidimensional problems.

Note that the algorithm (A.36) is quite efficiently realized in programming languages.

Appendix B

On modelling chemical reactions by the Monte Carlo method

Here algorithms of statistical modelling of chemical reactions are considered. A general scheme of modelling by using the direct Kolmogorov equation for homogeneous stepwise Markov processes is presented. The condition of coexistence of steady states in chemical systems is derived. An algorithm for calculation of quasi-potentials is given. Numerical examples are considered [27].

B.1. Introduction

Let volume V contain the molecules of M substances A_i, $i = 1, ..., M$. The molecules react with each other in N stages of a chemical reaction (under real conditions $V \sim 10^{23}$ molecules).

The chemical reaction can be represented in a general form [76] as:

$$\sum_{i=1}^{M} \alpha_{ji} A_i \rightleftarrows \sum_{i=1}^{M} \beta_{ji} A_i, \qquad j = 1, ..., N \tag{B.1}$$

i.e. α_{j1} molecules of the substance A_1, α_{j2} molecules of the substance $A_2, ..., \alpha_{jM}$ molecules of the substance A_M react to form β_{j1} molecules of the substance $A_1, ..., \beta_{jM}$ molecules of the substance $A_M, j = 1, ..., N$, and on the contrary, β_{j1} molecules of the substance $A_1, ..., \beta_{jM}$ molecules of the substance A_M react to form α_{j1} molecules of the substance $A_1, ..., \alpha_{jM}$ molecules of the substance A_M.

The coefficients α_{ji} and β_{ji} are assumed to satisfy the inequalities:

$$\sum_{i=1}^{M} \alpha_{ji} \leq 3, \quad \sum_{i=1}^{M} \beta_{ji} \leq 3, \quad j = 1, ..., N, \quad \alpha_{ji} \geq 0, \quad \beta_{ji} \geq 0.$$

Henceforth, we shall refer to these relations as condition (A1).

Let $x(t) = (x_1(t), ..., x_M(t))$ be the vector of concentrations of the substances involved in reaction (B.1) at a time moment t. Using the law of mass action the evolution of the system can be described by the system of differential equations [36]

$$\dot{x} = b(x), \quad x(0) = x_0, \quad b(x) = (b_1(x), ..., b_M(x)) \tag{B.2}$$

where

$$b_i(x) = \sum_{j=1}^{N} (\beta_{ji} - \alpha_{ji})(f_j(x) - g_j(x))$$

and $f_j(x), g_j(x)$ are defined as:

$$f_j(x) = k_j^+ + \prod_{i=1}^{M} x_i^{\alpha_{ji}}/\alpha_{ji}!, \quad g_j(x) = k_j^- - \prod_{i=1}^{M} x_i^{\beta_{ji}}/\beta_{ji}!$$

where k^+ and k^- are the constants of the jth stage of reaction (B.1) proceeding in the direct and reverse order, respectively.

Equation (B.2) is a mathematical expression that refers to the macroscopic approach to kinetics.

In what follows we assume that $b(x)$ belonging to (B.2) satisfies the condition

$$|b(x) - b(y)| \leq L|x - y|, \quad x, y \in D \tag{B.3}$$

where L is a constant and $D \in R^M$ is the range of concentration change.

Lately, probabilistic models of chemical reactions have been intensively studied [43], [60].

Let $X^V(t)$ be a monoparametric family of continuous on the right homogeneous stepwise Markov processes whose state space is X^M (the M-dimensional vectors with integral non-negative components), i.e. $X^V(t) = (X_1^V(t), ..., X_M^V(t))$ is a vector random quantity, where $X_i^V(t)$ is the number of molecules of the substance A_i at a time moment t.

The following result is given in [74]: *let $x(s)$ be a solution to equation (B.2) at $s \leq t$; then if condition (B.3) is satisfied with a probability close to unity, for any $s < \infty, V \to \infty$*

$$V^{-1}X^V(s) = x(s) + O(V^{-1/2}). \tag{B.4}$$

Therefore, one can conclude that at finite s, the stochastic trajectory only negligibly deviates from the deterministic one. Apart from the case of finite V and t, we are interested in the behaviour of the stochastic system (B.2) at large values of V and t tending to infinity.

In Section B.1 we give a general scheme of chemical reaction modelling when using the direct Kolmogorov equation for homogeneous stepwise Markov processes. In Section B.2 we derive the condition of coexistence of steady states in chemical reactions. In Section B.3 an algorithm for calculation of quasi-potentials is presented. In Section B.4 we give numerical examples of the implementation of the algorithm for a particular chemical reaction.

B.2. General scheme of chemical reaction modelling by the Monte Carlo method

Let

$$\alpha_j = \sum_{i=1}^{M} \alpha_{ji}, \quad \beta_j = \sum_{i=1}^{M} \beta_{ji}, \quad j = 1, ..., N.$$

Let us assume that the transition probabilisties of the process $X^V(t)$ satisfy the following conditions:

(1) the probability for the jth stage of reaction (B.1) to proceed directly in the interval $(t, t + \Delta t)$ is directly proportional to the total number of different interactions between molecules at the corresponding stage of (B.1) and inversely proportional to $V^{\alpha_j - 1}$, i.e.

$$R_j^+(X)\Delta t = k_j^+(V^{\alpha_j-1})^{-1}\left[\prod_{i=1}^{M}\binom{x_i}{\alpha_{ji}}\right]\Delta t$$

and, accordingly, for the reaction proceeding in the reverse order we have

$$R_j^-(X)\Delta t = k_j^-(V^{\beta_j-1})^{-1}\left[\prod_{i=1}^{M}\binom{x_i}{\beta_{ji}}\right]\Delta t$$

where $X = X^V(t)$;

(2) the probability for more than one stage to proceed in the interval $(t, t+\Delta t)$ is $o(\Delta t)$;

(3) the probability for no changes to occur in the interval $(t, t+\Delta t)$ equals

$$1 - \sum_{j=1}^{N}(R_j^+ + R_j^-)\Delta t + o(\Delta t);$$

(4) the probability for a stage notsatisfying condition (A1) to proceed equals zero.

Let

$$P(X;t) = P(X_1, ..., X_M; t)$$

be the probability that the volume V contains X_1 molecules of the substance $A_1, ..., X_M$ molecules of the substance $A_M, X \in X^M$, at a time moment t.

Then under some constraints (for instance, at bounded R_j which is the case here, as $V < \infty$) the direct Kolmogorov equation [63] is valid:

$$\frac{\partial P(X;t)}{\partial t} = -\left(\sum_j R_j\right) P(X;t) + \sum_{X'} P(X',t)\left(\sum R_j(X')\right)\frac{R_{j(x)}(X')}{\sum R_j(X')} \tag{B.5}$$

where $X \in X^M, X' \in X^M \setminus X$, the summation being made with respect to all the stages of reaction (B.1) proceeding in the direct and reverse order, $j(x)$ is an ordinal number of the stage starting from which the process passes from the state X' into $X, \sum_j R_j(X)\Delta t + o(\Delta t)$ is the probability for the vector $X^V(t)$ to undergo random changes in the interval $(t, t+\Delta t)$ if $X^V(t) = X$, and $R_{j(x)}(X')/\sum_j R_j(X')$ is the conditional probability for $X^V(t)$ to take on the value X at the time moment $t+\Delta t$, i.e. $X^V(t+0) = X$ on condition that $X^V(t) = X'$ and, additionally, $X^V(t)$ is known to have undergone changes in the interval $(t, t+\Delta t)$.

Let us take

$$q(X) = \sum_j R_j(X), \qquad Q(X', X) = \frac{R_{j(x)}(X')}{\sum_j R_j(X')}$$

then (B.5) takes the form

$$\frac{\partial P(X;t)}{\partial t} = -q(X)P(X;t) + \sum_{X'} P(X',t)q(X')Q(X',X). \tag{B.6}$$

Now, using equation (B.6) we obtain the integral equation [32]

$$P(X;t) = \exp(-\dot{q}(X)t) + \int_0^t \sum_{X'} P(X';s)\exp(-q(X)(t-s))q(X)Q(X',X)ds. \tag{B.7}$$

Taking into account (B.6) and (B.7), we find the equation for the collision density [19], where a collision is interpreted as an instantaneous change in the concentration vector.

Let a collision take place at a time moment s, i.e. $X^V(s) = X'$. The expression for the transition density $k(X', s; X, t)$ can be found from the fact that the quantity $k(X', s; X, t)$ equals the average number of collisions in $(t, t+\Delta t)$ immediately following the collision at s. From (B.6) we have that $Q(X', X)$ is the probability of $X^V(t+0) = X$. Then on condition that $X^V(t+0) = X$, the conditional probability of a collision in the interval $(t, t+\Delta t)$ equals $\exp(-q(X)(t-s))q(X)\Delta t$. According to the product rule, the absolute probability of a collision in the interval $(t, t+\Delta t)$ equals $Q(X', X)\exp(-q(X)(t-s))q(X)\Delta t$. Dividing the latter expression by Δt we obtain:

$$k(X', s; X, t) = Q(X', X)\exp(-q(X)(t-s))q(X).$$

Since the density of the distribution of the first-collision point equals $q(X)\exp(-q(X)t)$ [19], we get the following equation for the collision density:

$$\varphi(X; t) = q(X)\exp(-q(X)t) + \int_0^t \sum_{X'} \varphi(X'; s)k(X', s; X, t)ds. \qquad \text{(B.8)}$$

Equation (B.8) can be derived directly from (B.7) by drawing a parallel with transport theory. The collision density is known to be

$$\varphi(X; t) = \sigma(X)\Phi(X; t),$$

where $\sigma(X)$ is the total cross-section of interaction of the particle flow with a medium (the flow attenuation coefficient), $\Phi(X; t)$ is the flow of particles (the radiation intensity) [19]. In our case $\sigma(X) = q(X), \Phi(X; t) = P(X; t)$. Multiplying both sides of (B.7) by $q(X)$ we obtain equation (B.8). The integral operator entering this equation is the Volterra one and its spectral radius equals zero, which entails the possibility of applying the Monte Carlo method for estimating the functionals of the form $J = (\varphi, h)$.

The use of the Monte Carlo method in solving equation (B.8) is usually restricted to finding the moments of the random quantity $X^V(t)$.

From the standpoint of the theory of statistical modelling, to consider the model of chemical reactions defined as a process of pure creation is fruitful. Let

$$\Delta = \begin{bmatrix} \beta_{11} - \alpha_{11} & \dots & \beta_{1M} - \alpha_{1M} \\ \dots & \dots & \dots \\ \beta_{N1} - \alpha_{N1} & \dots & \beta_{NM} - \alpha_{NM} \\ \alpha_{11} - \beta_{11} & \dots & \alpha_{1M} - \beta_{1M} \\ \dots & \dots & \dots \\ \alpha_{N1} - \beta_{N1} & \dots & \alpha_{NM} - \beta_{NM} \end{bmatrix}$$

be the matrix of increments for reaction (B.1).

Let us consider a $2N$-dimensional vector $Y^V(t) = (Y_1^V(t), \dots, Y_{2N}^V(t))$ whose ith component $Y_i^V(t)$ is the number of occurrences of the ith stage of the reaction in the interval $(0, t)$, $i = 1, \dots, 2N$.

One can easily see that

$$X^V(t) = X^V(0) + \Delta^T Y^V(t), \qquad Y^V(0) = 0,$$

where Δ^T is the transposed matrix. Let $Y^V(t)$ be a monoparametric family of continuous on the right homogeneous stepwise Markov processes with the state space Z^{2N} (the $2N$-dimensional vectors with integral non-negative components), i.e. $Y^V(t) = (Y_1^V(t), ..., Y_{2N}^V(t))$ is a vector random quantity and $Y_i^V(t)$, $i = 1, ..., 2N$, are defined above.

The transition probabilities of the process $Y^V(t)$ are determined from the dependence between $X^V(t)$ and $Y^V(t)$ and from conditions (1-4). In the interval $(0, t)$, $t < \infty$, the number of states is always finite, which follows from the finiteness of the volume V. Since the moments of $Y^V(t)$ are known, one can easily find the corresponding characteristics of the process $X^V(t)$. Similarly to the process $X^V(t)$, the equation of the following form is valid:

$$\varphi(Y;t) = q(Y)\exp(-q(Y)t) + \int_0^t \sum_{Y'} \varphi(Y';s)k(Y',s;Y,t)ds$$

where $Y \in Z^{2N}, Y' \in Z^{2N} \setminus Y$ and $\varphi(Y;t)$ is the collision density, $q(Y)$ and $Q(Y', Y)$ have the same meaning as in (B.8). Let us rewrite the equation in the form

$$\varphi(Y;t) = f(Y;t) + \int_0^t \sum_{Y'} \varphi(Y';s)k(Y',s;Y,t)ds$$

or $\varphi = f + K\varphi$, $\varphi \in L$, $K \in [L \to L]$; L is the Banach space of integrable functions.

Now let us consider the estimates of functionals of the form:

$$J_h = (\varphi, h) = \sum_{n=0}^{\infty} (K^n f, h).$$

Suppose we have to determine the average number of occurrences of the ith stage of reaction (B.1) in a domain $\tilde{Y}_1 = \tilde{Y} \times [0, t]$, $i = 1, ..., 2N$:

$$\mathrm{E}Y_i^V(t) = \int_0^t \sum_{Y'} \varphi(Y';s)h_i(Y')ds, \quad Y' \in \tilde{Y}.$$

Let us consider a random quantity

$$\xi_i = \sum_{n=1}^{k} h_i(Y(n)),$$

where k is the random ordinal number of the last collision before leaving the domain $\tilde{Y}_1, Y(n)$ is the point of the nth collision. In what follows we shall need the following result [19] (see also Section 1.3).

Theorem B.1. *If $\|K^{n_0}\| < 1$ at some natural n_0, then*

$$\mathrm{E}\xi_i = \mathrm{E}\sum_{n=1}^{k} h_i(Y(n)) = (\varphi, h_i) = \mathrm{E}Y_i^V(t), \quad \mathrm{D}\xi_i < +\infty. \tag{B.9}$$

The theorem entails the unbiased character of the following estimates. Let

$$h_i(Y(m)) = \begin{cases} 1, & \text{if the collision resulted in the } i\text{th stage of (A)} \\ 0, & \text{otherwise.} \end{cases}$$

Substituting $h_i(Y(m))$ into (B.9) we obtain the collision estimate for the number of occurrences of the ith stage of reaction (B.1), $i = 1, ..., 2N$. It is this estimate that is used in [29], [57]. If one sets

$$h_i(Y(m)) = R_i(Y(m))/\sum_i R_i(Y(m)),$$

the quantity $\xi_i = \sum_{m=1}^{k} h_i(Y(m))$ is an unbiased estimate by the expectation method [19]. While in using the collision estimate, one calculates the actual number of occurrences of the stage, in using the estimate by the expectation method, the expectation of the occurence of a reaction stage is fixed at each collision point.

Assume that $h_i(Y(m)) = R_i(Y(m))\tau_m$, where τ_m is the free path between the $(m-1)$th and the mth collisions. Then the quantity

$$\xi_i = \sum_{m=1}^{k} h_i(Y(m))$$

is the unbiased path estimate with the finite variance (see Section 3.6).

Let us consider the following chemical reaction

$$1) X + Y \xrightarrow{k_1} 2X, \qquad 2) X \xrightarrow{k_2} Y$$

where $k_1 = k_2 = 1$. Let us denote the concentrations of the reactants X and Y by x_1 and x_2, respectively. Making use of the law of mass action we have the system of differential equations:

$$\dot{x}_1 = x_1 y_1 - x_1 \qquad \dot{y}_1 = -x_1 y_1 + x_1 .$$

Let us set $x_1(0) = 0.2, y_1(0) = 0.8$. Taking into account the balance requirement $x_1+y_1 = 1$ we obtain $\dot{x}_1 = -x_1^2$. Solving this equation we find

$$x_1 = 0.2/(1 + 0.2t), \qquad y_1 = (0.8 + 0.2t)/(1 + 0.2t).$$

Now let us turn to the stochastic analogue. Let us set the system size $V = 10000$. Let $X_1(t), X_2(t)$ be the numbers of molecules of the substances X and Y at a time moment t. Using conditions (1-4) we determine the transition probabilities of the embedded Markov chain:

$$P((X_1, X_2) \to (X_1 + 1, X_2 - 1)) = \frac{0.0001 X_1 X_2}{0.0001 X_1 X_2 + X_1}$$

$$P((X_1, X_2) \to (X_1 - 1, X_2 + 1)) = \frac{X_1}{0.0001 X_1 X_2 + X_1}.$$

The free path between collisions is

$$\tau = -\frac{\ln \alpha}{0.0001 X_1 X_2 + X_1}$$

where α is a random number uniformly distributed over the interval $[0, 1]$. The initial conditions take on the form $x_1(0) = 2000, x_2(0) = 8000$.

For the estimate by the expectation method we have:

$$h_1 = \frac{0.0001X_1X_2}{0.0001X_1X_2 + X_1}$$

$$h_2 = \frac{X_1}{0.0001X_1X_2 + X_1}$$

And for the path estimate we obtain $h_1 = 0.0001X_1X_2\tau, h_2 = X_1\tau$. In this case

$$\Delta = \begin{bmatrix} 1 & -1 \\ -1 & 1 \end{bmatrix}$$

and for the increments in concentrations we have the following expression:

$$\Delta X = \Delta^T \begin{bmatrix} h_1 \\ h_2 \end{bmatrix}.$$

Numerical calculations carried out for the number of realizations $k = 10$ are in good agreement with the exact solution. With the same time cost the expectation method and path estimates prove to be more efficient than the 'collision' ones.

B.3. Conditions of coexistence of steady states in chemical systems

In studying chemical reactions it is important to know with probability close to unity what behaviour of the process $X^V(t)$ is to be expected in a large time interval. Thus, finding the asymptotic probabilistic estimates of large deviations of the process $X^V(t)$ in a finite time interval $[0, T]$ is of particular significance, i.e. estimating the probabilities $P(X^V(t) \in A)$, $t \in [0, T]$ for the set A of the functions that do not contain the dynamic trajectory $x(t)$ corresponding to $X^V(t)$.

If the function $x(t)$ along with some of its neighbourhood does not belong to the set A, then $P(X^V(t) \in A) \to A$, $V \to \infty$.

In this case one can specify the function $\varphi \in A$ such that the major part of the probabilistic measure of the set A is concentrated near the function φ, i.e. for any neighbourhood $U(\varphi)$ of this function

$$P(X^V(t) \in A \setminus U(\varphi)) = o(P(X^V(t) \in U(\varphi)))$$

at $V \to \infty$. It turns out that one can introduce the functional $S(\varphi)$ of functions $\varphi \in A$ defined in the interval $[0, T]$, such that at sufficiently small δ and at $V \to \infty$

$$P(\rho(X^V(t), \varphi(t)) < \delta) \approx \exp(-VS(\varphi))$$

where ρ is the distance in the space of the functions conditions in the interval $[0, T]$ [74]. If the minimum of the functional $S(\varphi)$ on the set A is attained for the function $\tilde{\varphi}$, it is clear that at large V the main contribution to the probability $P(X^V(t) \in A)$ will be made by the neighbourhood of the function $\tilde{\varphi}$.

By analogy with classical mechanics, $S(\varphi)$ is usually called the action functional. Below, the action functional for the process $X^V(t)$ is given. We assume that for all $\alpha = (\alpha_1, ..., \alpha_M)$ the expression

$$H(x,\alpha) = \sum_i \beta_i\alpha_i + \int_{R^M} (\exp\{\sum \alpha_i\beta_i\} - 1 - \sum \alpha_i\beta_i)\mu_x \mathrm{d}\beta$$

is finite, where μ_x at every $x \in R^M$ is the measure of discontinuities on $R^M \setminus \{0\}$ such that

$$\int_{R^M\setminus\{0\}} |\beta^2|\mu_x \mathrm{d}\beta < \infty.$$

Let us introduce the notation $L(x,\beta)$ for the Legendre transformation of the function $H(x,\alpha)$ over the second argument, i.e.

$$L(x,\beta) = \sup_\alpha |(\alpha,\beta) - H(x,\alpha)|.$$

Let us consider the function $\varphi(t), T_1 \leq t \leq T_2$, with the values in R^M and define for this function the functional:

$$S(\varphi) = S_{T_1T_2}(\varphi) = \int_{T_1}^{T_2} L(\varphi,\dot{\varphi})\mathrm{d}t$$

if the function φ is absolutely continuous and the integral converges; otherwise $S(\varphi) = \infty$. It is this functional that will be the normalized action functional (the normalizing factor will be V)[74].

Let a point $O \in R^M$ be an asymptotically stable position of equilibrium of (B.2). The quasi-potential of dynamic system (B.2) with respect to the point O is a function $V(O,x)$ defined by the equality:

$$V(O,x) = \inf\{S(\varphi) : \varphi \in C_{T_1T_2}(R^M), \quad \varphi(T_1) = 0, \quad \varphi(T_2) = x\}.$$

The endpoints T_1 and T_2 of the interval are not fixed here.

Now we pass directly to deriving the condition of coexistence of steady states in chemical systems. Without loss of generality we shall consider a non-dimensional chemical reaction and put it into correspondence with the Markov process $X_k^V(t)$ that for each $k > 0$ meets the following conditions analogous to conditions (1-4): the process starting from the state X in time Δt jumps into the state $X+1$ with the probability $B(X,k)\Delta t$, into the state $X-1$ with the probability $E(X,k)\Delta t$ and remains in the same state with the probability $1-(B(X,k)+E(X,k))\Delta t + o(\Delta t)$. Equation (B.4) implies that at large V the dynamics of the process is described by the differential equation:

$$\dot{x} = r(x,k) - l(x,k) \tag{B.10}$$

where

$$r(x,k) = B(X,k)/V + O(1/V), \quad l(x,k) = E(X,k)/V + O(1/V).$$

Suppose that the domain of the parameter k is partioned into two subdomains:

$$K_1 = \{k : 0 \leq k \leq k_1\} \cup \{k : k \geq k_2\}, \quad K_2 = \{k : k_1 \leq k \leq k_2\}.$$

Let equation (B.10) possess a unique asymptotically stable position of equilibrium $x_{st}(k)$ for any $k \in K_1$ and three positions of equilibrium $x_1(k), x_2(k), x_3(k)$ for any $k \in K_2$, where

$x_1(k) < x_2(k) < x_3(k)$, $x_1(k)$ and $x_3(k)$ are asymptotically stable equilibrim positions, $x_2(k)$ is an unstable equilibrium position.

For $k \in K_1$ and $x \in X$, the trajectory of equation (B.10) originating at the point x tends to x_{st} at $t \to \infty$.

For any $k \in K_2$, we obtain two nonintersecting domains

$$D_1 = \{x : 0 \leq x \leq x_2\}, \qquad D_2 = \{x : x > x_2\}.$$

The trajectories originating at the points $x \in D_1(D_2)$ tend to $x_1(x_3)$ at $t \to \infty$. There are experimental examples [42] when the jump from one branch of steady states to the other and vice versa occurs at some internal point $k_3, k_1 < k_3 < k_2$. This is similar to the first-order phase transitions in statistical physics. Thus, one must predict the value of parameter k, at which the jump from one branch to the other takes place.

The problem of establishing the conditions of coexistence of two steady modes is naturally linked with the problem of the limit behaviour of the steady distribution $\mu_k^{1/V}$ of the process $X_k^V(t)$ at $V \to \infty$.

Taking into account (B.4) we establish the correspondence between the solutions to equation (B.10) and the process $X_k^V(t)$, for example, $X_k^V(t) = [xV]$, where $[\cdot]$ stands for the operation of taking the integral part. It is clear that the process $X_k^V(t)$ proceeds most of the time in the neighbourhood of stable equilibrium positions of equation (B.10). From time to time it departs by appreciable distance from $X_1(k)$ or $X_3(k)$ and returns again to their neighbourhood. And very rarely the process passes from the neighbourhood of $X_1(k)$ to the neighbourhood of $X_3(k)$ and vice versa.

Let us introduce the following quantites:

$$U_1(x_i(k), x_j(k)) = \inf\{S_{0T}(\varphi) : \varphi(0) = x_i(k), \varphi(T) = x_j(k)\}$$

where $i, j = 1, 2, 3, i \neq j$, and

$$\tilde{U}(x_i(k), x_j(k)) = \inf\{S_{0T}(\varphi) : \varphi(0) = x_i(k), \varphi(T) = x_j(k), \varphi(T) \in D \setminus \cup_{s \neq i} x_s(k)\}.$$

Then one can show that

$$U_1(x_1(k), x_3(k)) = \tilde{U}(x_1(k), x_2(k)) + \tilde{U}(x_2(k), x_3(k))$$

$$U_1(x_3(k), x_1(k)) = \tilde{U}(x_3(k), x_2(k)) + \tilde{U}(x_2(k), x_1(k)).$$

If $X_2(k)$ is an unstable equilibrium position, the following equalities are valid [74]:

$$\tilde{U}(x_2(k), x_1(k)) = 0, \qquad \tilde{U}(x_2(k), x_3(k)) = 0.$$

The latter sequence of equalities can be rewritten in the following form:

$$U_1(x_1(k), x_3(k)) = \tilde{U}(x_1(k), x_2(k)) = U_{12}, \quad U_1(x_3(k), x_1(k)) = \tilde{U}(x_3(k), x_2(k)) = U_{32}.$$

Let us cite some results from graph theory. Let L be a finite set whose elements are denoted by letters i, j, k, m and so on. Now let us choose a subspace W of this set. By the W-graph we shall denote the graph consisting of the arrows $m \to n$ ($m \in L \setminus W, n \in L, m \neq n$), provided it meets the following requirements:

(1) exactly one arrow originates at each point $m \in L \setminus W$;

(2) there are no closed cycles in the graph.

The set of all the W-graphs is denoted by $G(W)$, the letter g being used for the notation of separate graphs.

To find the asymptotics of the steady distribution the values of the following quantities are essentially needed:

$$W(x_i(k)) = \min_{g \in G(W)} \sum_{(m \to n) \in g} \tilde{U}(x_m(k), x_n(k)). \tag{B.11}$$

Using (B.11) for the stable equilibrium positions of our system we obtain:

$$W(x_1(k)) = \tilde{U}(x_3(k), x_2(k)) + \tilde{U}(x_2(k), x_1(k)) = U_{32},$$

similarly,

$$W(x_3(k)) = \tilde{U}(x_1(k), x_2(k)) + \tilde{U}(x_2(k), x_3(k)) = U_{12}.$$

On the other hand, it is known that for unstable equilibrium positions

$$W(x_2(k)) = \min[W(x_i(k)) + \tilde{U}(x_i(k), x_2(k))]$$

where the minimum is taken over all the stable equilibrium positions [74].

From (B.11) it follows that the minimum $W(x_i)$ may be attained only in the stable equilibrium positions. Now we apply the following theorem from [74].

Theorem B.2. *Let us take*

$$W(x) = \min[W(x_i(k)) + U_1(x_i(k), x)] \tag{B.12}$$

for $x \in D$, where the minimum may be taken either over all the equilibrium positions or only over the stable ones. Let γ be any positive number. Then for any sufficiently small neighbourhood $E_\rho(x)$ of the point x, there exists $E_0 > 0$ such that at $E < E_0$ we have:

$$\exp\{-V(W(x) - \min_i W(x_i(k)) + \gamma)\} \leq \mu^{1/V}(E_\rho(x))$$
$$\leq \exp\{-V(W(x) - \min_i W(x_i(k)) - \gamma)\}.$$

This theorem implies that the asymptotics of the steady distribution $\mu^{1/V}$ at $V \to \infty$ is given by the action function:

$$V(W(x) - \min W(x_i(k))).$$

Using (B.11)-(B.12) we can easily give a geometrical interpretation of Theorem B.2. Indeed, at $x = x_1(k)$ we have

$$W(x_1(k)) - \min W(x_i(k)) = \tilde{U}(x_3(k), x_2(k)) - \min W(x_i(k)), \tag{B.13}$$

similary, at $x = x_3(k)$ we obtain

$$W(x_3(k)) - \min W(x_i(k)) = \tilde{U}(x_1(k), x_2(k)) - \min W(x_i(k)). \tag{B.14}$$

On the other hand, if

$$\min W(x_i(k)) = \tilde{U}(x_3(k), x_2(k)) < \tilde{U}(x_1(k), x_2(k))$$

then, taking into account the theorem and (B.13), (B.14) we see that at $V \to \infty$ the steady distribution is concentrated in the vicinity of the point $x_1(k)$ and, on the contrary, if

$$\min W(x_i(k)) = \tilde{U}(x_1(k), x_2(k)) < \tilde{U}(x_3(k), x_2(k))$$

the steady distribution is concentrated in the vicinity of $x_3(k)$. Thus, of interest is the case when $U_{32} = U_{12}$. In such a situation differences (B.13), (B.14) are equal to 0 and one may assume that two steady states of the system coexist.

This condition (the equality of quasi-potentials taken with respect to the stable equilibrium positions) is similar to the Maxwell rule which states that chemical potentials at the point of phase transition of the type 'water-vapour', etc. are equal.

B.4. Calculation of quasi-potentials of dynamic systems

Let a point $O \in R^M$ be an asymptotically stable position of equilibrium of (B.2), i.e. for any neighbourhood E_1 of the point O one can find a smaller neighbourhood E_2 such that the trajectory of the system (B.2) originating in E_2 tends to O without leaving the neighbourhood E_1 at $t \to \infty$.

We say that the domain D is attracted to the point O if the trajectories $x_t(x), x \in D$, tend to the equilibrium position O at $t \to \infty$ without leaving the domain D.

Suppose that for $x \in \partial D$, we have $(b(x), n(x)) < 0$, where ∂D is the boundary of the domain $D, n(x)$ is the exterior normal to the boundary of the domain D. From [74] it follows that at large V, the following statements are valid for the process $X^V(t)$.

(1) If there exists only one point $Y_0 \in \partial D$ such that $V(0, Y_0) = \min_{Y \in \partial D} V(O, Y)$, then for any $\delta > 0, x \in D:$

$$\lim_{V \to \infty} P_X\{\rho(X^V(\tau_V), Y_0) \leq \delta\} = 1 \tag{B.15}$$

where

$$\tau_V = \inf\{t : X^V(t) \in \partial D\}.$$

(2) For any point $x \in D$, the Markov trajectories $X^V(t)$ starting from this point will fall into a small neighbourhood of the equilibrium position before they reach the boundary ∂D with the probability tending to unity at $V \to \infty$.

(3) At $x \in D$ we have:

$$\lim_{V \to \infty} V^{-1} \ln M_X \tau_V = V_0 = \min_{Y \in \partial D} V(O, Y). \tag{B.16}$$

Besides, we assume that any trajectory of system (B.2) originating at any point $x \in D \cup \partial D$ reaches a small neighbourhood of the equilibrium position in finite time.

For large V, the correspondence between the process $X^V(t)$ and the trajectory $X(t)$ is established in accord with (B.4): $X^V(t) = [xV]$, where $[\cdot]$ stands for the integral part. The h-jumps of the process $X^V(t)$ are supposed to equal ± 1. If $h \neq \pm 1$ the procedure suggested below is to be slightly modified.

Without loss of generality the algorithm will be constructed for the domain $D \in R^2$. Let us suppose that we have to calculate the value of the quasi-potential at the point $x \in D$. Let us pass to the process $X^V(t)$. If $X^V(0) = O$, then at large V for any finite t, the deviations of $X^V(t)$ from the equilibrium position are $O(V^{1/2})$, i.e. they are negligible in terms of concentations. Thus, the problem of calculating the value of the quasi-potential at a point rather distant from the equilibrium position cannot even be posed if one adheres to the direct modelling approach. Therefore, we shall construct the process $X_1^V(t)$ which in its outline imitates $X^V(t)$ and is more effective for implementing on a computer.

Let us fix a number V (usually, acceptable results are already obtained at the values as small as $V \sim 10^3$). Further, we consider a discrete model and define the process $X^V(t)$.

Let R_1 be the distance between the points O and X. Let us set time T and let

$$d = \sup_{t<T} |X^V(t) - X^V(0)|$$

be determined from a numerical experiment and let $R = R_1 - d$.

The small neighbourhood of the equilibrium position is defined as follows: it is taken to be a circle γ_1 of radius $d/2$. Let us construct a system of concentric circles centred at the point X with the radii $r_k = R - k, k = 0, ..., R - 1$. Here we take into account that the jumps of the process equal $h = \pm 1$. Domain D_k coincides with the exterior of the circle of radius r_k.

Let the process $X^V(t)$ start from the point O. Let us trace its trajectory till the moment of its first fall into the domain $K_1 = \partial D_0 \cup D_1 \setminus D_0$ and fix the obtained point O_1. If the process proceeding from the point O_1 visits the interior of the domain

$$K_m = \partial D_{m-1} \cup D_m \setminus D_{m-1}, m = 2, ..., R - 1,$$

before it for the first time gets into γ_1, we shall fix the point O_m and follow the trajectory of the process till it falls inside γ_1. After the process has fallen into the domain γ_1 the next realization is taken to start from the point O, all the states of the process from the domain γ_1 being considered as a single aggregate state. At large V, the time spent on random walks from the point O before getting into γ_1 coincides with the time t_m of the evolution of dynamic system (B.2).

Thus, we have arrived at the following procedure: the process starts from the point O, reaches some point O_m closest to X in some time τ_m and then returns to the equilibrium position in the time t_m.

The neighbourhood γ_1 of the equilibrium position is transformed into the neighbourhood γ_m which is a circle centred at the point O with the radius $r = d/2 + m$, since, according to statement (2), having fallen inside γ_m, the process gets into the neighbourhood γ_1 (and, hence, automatically into the point O) with probability close to unity before arriving at the boundary of the domain D_m.

According to statement (B.15), the trajectory of the process $X^V(t)$ starting from the point O will arrive at the boundary of the domain K_m at the point close to O_m (at large V they become indistinguishable) with probability approaching unity.

Therefore, we assume that the process restarting from the point O in some random time interval τ_m will fall inside a small neighbourhood of the point O. The times $\tau_m(l)$ thus defined are independent owing to the process being of the Markov type. They are considered to be identically distributed.

Let η_{mn} be a random quantity equal to the number of restorations that occurred until the process starting from the point O_m gets inside the domain $K_n = \partial D_{n-1} \cup \partial D_n \setminus D_{n-1}$.

Now we fix a given point O_n, construct the corresponding domain γ_n and repeat the previous procedure. Ultimately, we shall get inside the K_{R-1} neighbourhood of the point X while moving along the system of concentric circles.

It is the broken-line trajectory marked with the points $O, O_1, ..., O_{R-1}, X$ that approximately represents the extremal connecting the points O and X. Different computer realizations, generally speaking, produce different broken lines. But they differ only slightly owing to the uniqueness of the extremal [74]. The process $X_1^V(t)$ obtained imitates the process $X^V(t)$ and can easily be implemented on a computer. The expectation of the time required for the process $X_1^V(t)$, which starts from the equilibrium position, to reach the point X can be found from the recurrent relationships:

$$M\tau_2 = (M\tau_1 + t_1)(1 + M\eta_{12}), ..., M\tau_R = (M\tau_{R-1} + t_{R-1})(1 + \eta_{R-1,R}).$$

Then the value of the quasi-potential of the field $b(x)$ with respect to the point O is

$$V(O, X) \approx \ln M\tau_R / V.$$

To check out the validity of the algorithm we have considered some examples.

B.5. Examples

Let us first consider an example for which the exact results are known. In [10], the following reaction

$$1) Z \underset{k_{-1}}{\overset{k_1}{\rightleftarrows}} X \qquad 2) X + 2Z \xrightarrow{k_2} 3Z$$

was subjected to deterministic analysis. Let us denote the concentration of the intermediate substance Z by z. The dynamic equation for this reaction takes the form:

$$\dot{z} = r(z) - l(z)$$

where

$$r(z) = (k_{-1} + k_2 z^2)(1 - z), \quad l(z) = k_1 z.$$

The quasi-potential of the field $r(z) - l(z)$ with respect to the equilibrium position z_0 is determined from formula [74]:

$$V(z_0, z) = \int_{z_0}^{z} \ln \frac{l(z)}{r(z)} dz.$$

Taking into account the expression for $r(z)$ and $l(z)$ we obtain:

$$\begin{aligned} V(z_0, z) &= (2 + z \ln 1.78) z + (1 - z) \ln(1 - z) \\ &- z \ln(k_{-1} + k_2 z^2) - 2(k_{-1}/k_2)^{1/2} \text{arctg}(k_2/k_{-1})^{1/2} + C \end{aligned}$$

where C is a constant dependent on the lower integration limit.

Let us set $k_1 = k_2 = 1$ and $k_{-1} = 1.74$. Then $z_0 = 0.4$ and

$$\begin{aligned} V(z_0,z) &= (2+z\ln 1.78)z+(1-z)\ln(1-z)-z\ln(1+z^2)-2\,\mathrm{arctg}\,z \\ &- (2+0.4\ln 1.78)-0.6\ln 0.6+0.4\ln 1.16+2\,\mathrm{arctg}\,0.4. \end{aligned} \tag{B.17}$$

Suppose that we have to find the value of the quasi-potential at the point $z = 0.8$. According to (B.17) we have $V(0.4; 0.8) \approx 0.14$.

Now let us turn to the numerical experiment and take the volume $V = 500$. We have to find the value of the quasi-potential at the point $Z = 400$, if the equilibrium position is $Z_0 = 200$. In the one-dimensional case the algorithm is significantly simpler, as there is only one trajectory connecting the points Z_0 and Z.

For $n = 100$ realizations, the value $V(200; 400) = 1.39 \pm 0.02$ was obtained, the time of computation on BESM - 6 computer being $t = 2\,\mathrm{min}$.

Now let us consider the two-dimensional model of enzymolytic reaction [73]:

$$1) A + X \underset{k_{-1}}{\overset{k_1}{\rightleftarrows}} 2X \qquad 2) X + E \underset{k_{-2}}{\overset{k_2}{\rightleftarrows}} C \qquad 3) C \underset{k_{-3}}{\overset{k_3}{\rightleftarrows}} B + E.$$

Here A and B are the primary and end products of the reaction. They are assumed to be constant. E is a free enzyme. The relationship $C + E = E_T$ is valid for C and E. Let us write down the system of differential equations describing the evolution of concentrations of x and E:

$$\begin{aligned} \dot{x} &= k_1Ax - k_{-1}x^2 - k_2xE + k_{-2}(E_T - E) \\ \dot{E} &= -k_2xE - k_{-3}BE + (k_{-2} + k_3)(E_T - E). \end{aligned} \tag{B.18}$$

Let us take

$$k_1A = 3, \;\; k_{-1} = 1, \;\; k_2 = 1, \;\; k_{-2} = 1, \;\; k_3 = 1, \;\; k_{-3}B = 1, \;\; E_T = 9.$$

The equilibrium position O has the coordinates $x_{st} = 2.08$, $y_{st} = 3.54$. Suppose that we have to determine the value of the quasi-potential at a point with the coordinates $x_F = 0$, $E_F = 3$. There are no explicit expressions for the quasi-potential in the two-dimensional case. Therefore, let us proceed directly to numerical experiments. Let the volume $V = 500$, the equilibrium position have the coordinates $X_{st} = 1040$, $E_{st} = 1770$. We have to find the value of the quasi-potential at the point $X = 0$, $E = 1500$. The distance between these points equals $R_1 = 1074$.

The maximal deviation of the process from the equilibrium position obtained from the numerical experiments is $d = 70$, thus, $R = 1004$. We construct a system of circles centred at the point F with the radii $r_k = R - k, k = 0, ..., R - 1$, as the jumps of the process are $h = \pm 1$.

Now we come into the neighbourhood of the point F along the system of circles. For the number of realizations $n = 10$ the obtained value of the quasi-potential was $V(O; F) = 2.573 \pm 0.03$. The computing time on BESM-6 computer is $t = 8\,\mathrm{min}$.

In [27] they proposed the following expression for the quasi-potential of the reaction under study:

$$V(O; F) = \int\limits_L \left(\ln \frac{k_1Ax + k_{-2}(E_T - E)}{k_{-1}x^2 + k_2xE}\mathrm{d}x + \ln \frac{(k_{-2} + k_3)(E_T - T)}{k_2xE + k_{-3}BE}\mathrm{d}E \right) \tag{B.19}$$

the integral being taken along the trajectory of the dynamic system

$$\dot{\varphi}_1 = k_1 A\varphi_1 + k_{-2}(E_T - \varphi_2) - k_{-1}\varphi_1^2 - k_2\varphi_1\varphi_2$$

$$\dot{\varphi}_2 = (k_{-2} + k_3)(E_T - \varphi_2) - k_2\varphi_1\varphi_2 - k_{-3}B\varphi_2$$

that connects the small neighbourhood of the point O with the point F. The value of the quasi-potential obtained is $V(O; F) = 2.542$, which is very close to that obtained from the numerical experiment.

In [27] they consider system (B.18) with the following values of parameters:

$$k_1 = k_{-1} = 1, \quad k_2 = k_3 = k_{-3} = 1, \quad k_{-2} = 0.35, \quad B = 0.1, \quad E_T = 1.$$

At $A \in (2.41; 2.67)$ system (B.18) has three equilibrium positions: two asymptotically stable of the 'node' type and one of the 'saddle' type. The critical value A_c of the parameter A at which the phase transition occurs is found from (B.19): $A_c = 2.505$. The Monte-Carlo analysis based on the ergodic Birkhoff-Hynchin theorem gives $A_c = 2.51$ for $V = 40$. By convention we call our algorithm the 'mountaineer' algorithm. Being applied to the system with $V = 500$ it yields the value $A_c = 2.53$. Because of the many-fold increase in the size of the system, it is the latter result that should be considered as the one with the least deviation from the exact value.

The algorithm considered can easily be applied to chemical reactions of arbitrary dimension. However, this may result in a significant increase of the computing time, the illustration being one and two-dimensional examples in the present section. The qualitative analysis of the systems of type (B.2) also becomes much more complicated.

Appendix C
One unsolved minimax problem

Monte Carlo estimates for integrals are constructed on the basis of the relations

$$I_k = \int_X f_k(x)\mathrm{d}x = E\zeta_k, \quad \zeta_k = f_k(\xi)/p(\xi), \quad k = 1, 2, \ldots, n,$$

where $p(x)$ is the distribution density for ξ, and

$$D\zeta_k(p) = \int_X \frac{f_k^2(x)\mathrm{d}x}{p(x)} - I_k^2.$$

Let

$$F_\zeta(p) = \max_k\{D\zeta_k(p)\}.$$

We consider the following problems: find density $p = p^*(x)$ such that $\min_p F_\zeta(p) = F_\zeta(p^*)$ is attained. We formulate the known [75] relation between the minimax and the Bayes solutions in the form of the following statement.

Lemma C.1. *Let $\xi_k(p)$ be random variables in a probabilistic model P of a class of admissible models, and let P_Λ be a model such that the minimum*

$$\min_P \sum_k \lambda_k D\xi_k(p) = G(\Lambda), \quad 0 \le \lambda_k < +\infty, \quad \Lambda = (\lambda_1, \ldots, \lambda_n),$$

is attained, where $G(\Lambda) < \infty$ and $D\xi_k(P_\Lambda)$ is differentiable with respect to λ_i $(i, k = 1, \ldots, n)$ for all Λ. Then $\min F_\xi(p)$ is attained in P_{Λ_0}, where Λ_0 is the solution to the problem

$$\max_\Lambda \left\{ \sum_{k=1}^n \lambda_k D\xi_k(P_\Lambda) : \ \sum_{k=1}^n \lambda_k = 1 \right\}.$$

The proof of the lemma consists in justifying the fact that the variances are equal on the support of the measure of Λ_0 and that the others do not exceed this value [49]. Then, the standard inequality [75] is applied.

Note that Lemma C.1 cannot be directly derived from [75], where the set of solutions $\{P\}$ was assumed to be compact. Formally, it can be considered as a special case of the minimax theorem [75] because

$$\max_k D\xi_k = \max_\Lambda \left\{ \sum_{k=1}^n \lambda_k D\xi_k : \ \sum_k \lambda_k = 1, \ \lambda_k \ge 0 \right\}.$$

To obtain the required result, it is sufficient to invert the order of min and max. However, in the standard minimax theorem, the set $\{P\}$ is also assumed to be compact; therefore, a simple proof of Lemma C.1 is reasonable.

Note that the solution to the problem of optimization of estimates for integrals according to the criterion based on the weighted sum of the variance is well-known and is given by the relations

$$\min_p \sum_{k=1}^{n} \lambda_k D\xi_k(p) = \sum_{k=1}^{n} \lambda_k D\xi_k(P_\Lambda) = G_n(\Lambda),$$

where

$$P_\Lambda(x) = c\Big[\sum_{k=1}^{n} \lambda_k f_k^2(x)\Big]^{1/2}, \qquad c^{-1} = \int \Big[\sum_{k=1}^{n} \lambda_k f_k^2(x)\Big]^{1/2} \mathrm{d}x,$$

$$G_n(\Lambda) = \Big\{\int \Big[\sum_{k=1}^{n} \lambda_k f_k^2(x)\Big]^{1/2} \mathrm{d}x\Big\}^2 - \sum_{k=1}^{n} \lambda_k I_k^2.$$

Here the nature of the space X and of the integration measure is not important. However, we assume for simplicity that X is a finite-dimensional Euclidean space, and the integrals are taken with respect to the Lebesgue measure in X. Below, we consider the statements concerning the minimax of the variances of estimates of n integrals (see [49]).

Lemma C.2. *Let*

$$P_0 = \int \sup_k |f_k(x)| \mathrm{d}x < +\infty.$$

Then the optimal density is given by

$$p^*(x) = p_{\Lambda^*}(x),$$

where Λ^* is a maximum point for $G_n(\Lambda)$.

Consider also the case where $f_\sigma = f(x, \sigma)$ depends continuously on σ, $\sigma \in [\sigma^{(1)}, \sigma^{(2)}]$. Introduce the following notation: Λ is a probabilistic measure on $[\sigma^{(1)}, \sigma^{(2)}]$, $\zeta_\sigma(p) = \zeta(p, \sigma)$

$$P_0 = \int \Big[\sup_\sigma |f(x,\sigma)|\Big] \mathrm{d}x, \quad \varphi^2(x) = \inf_k \int f^2(x,\sigma)\Lambda_k^*(\mathrm{d}\sigma),$$

where $\Lambda_k^* = (\lambda_1^*, \dots, \lambda_{k+1}^*)$ is the optimal (in the sense of Lemma C.2) discrete measure for the set of values of σ coinciding with the $k+1$ nodes of the uniform network of $[\sigma^{(1)}, \sigma^{(2)}]$.

Lemma C.3. *Suppose $P_0 < +\infty$, $|f(x,\sigma) f_\sigma'(x,\sigma)| \le h(x)$,*

$$\int \frac{f^2(x,\sigma)}{\varphi(x)} \mathrm{d}x < c_1 < +\infty, \quad \int \frac{h(x)}{\varphi(x)} \mathrm{d}x < c_2 < +\infty.$$

Then,

$$p^*(x) = p(x, \Lambda^*) = c\Big[\int f^2(x,\sigma)\Lambda^*(\mathrm{d}\sigma)\Big]^{1/2},$$

where Λ^* is a maximum point for

$$G(\Lambda) = \Big\{\int \Big[\int f^2(x,\sigma)\Lambda(\mathrm{d}\sigma)\Big]^{1/2} \mathrm{d}x\Big\}^2 - \int I^2(\sigma)\Lambda(\mathrm{d}\sigma).$$

Proof, see in [49].

A series of examples has shown that sometimes it is useful to take

$$p \equiv p_0(x) = P_0^{-1} \sup_{\sigma} |g(x,\sigma)| \Rightarrow D\zeta(p_0,\sigma) \le I_1(\sigma)[P_0 - I^2(\sigma)/I_1].$$

If the relative variances $D\zeta(p,\sigma)/I^2(\sigma)$ are to be minimized, then $g(x,\sigma)$ is replaced by

$$\frac{|g(x,\sigma)|}{I_1(\sigma)}, \qquad I_1(\sigma) = \int_X |g(x,\sigma)|\, dx.$$

If two integrals of non-negative functions g_1, g_2 are calculated, it is then useful to take the density

$$p^{(0)}(x) = \frac{g_1(x) + g_2(x)}{I_1 + I_2}.$$

Now,

$$\zeta_2 = (\zeta_1 + \zeta_2) - \zeta_1, \qquad D[\zeta_1(p^{(0)}) + \zeta_2(p^{(0)})] = 0;$$

therefore $D\zeta_1(p^{(0)}) = D\zeta_2(p^{(0)})$. Consequently, if $g_1 g_2 \equiv 0$, then $p^* = p^{(0)}$, since in this case $g_1^2 + g_2^2 \equiv (g_1 + g_2)^2$. On the other hand, $p^{(0)}$ is optimal if $g_1 \equiv g_2$. So it is hoped that $p^{(0)}$ will give a good result in a broad class of cases ([49], Sect. 4.3). It is not difficult to see that the relative variances are equal if $p_1^{(0)} = g_1/I_1 + g_2/I_2$.

Consider a case [49] important for the minimax theory of the weight Monte Carlo algorithms

$$f(x,\sigma) = \sigma \exp(-\sigma x), \quad X = [0, +\infty), \quad 0 < \sigma_1 < \sigma_2 < \infty.$$

Calculations carried out for a large set of values of σ_1 and σ_2 show that the optimal density p^* coincides in this case with the density p_{12}^* obtained by changing the segment $[\sigma_1, \sigma_2]$ with the two-element set $\{\sigma_1, \sigma_2\}$, and has the form

$$p_{12}^*(x) = c[\lambda \sigma_1^2 \exp(-2\sigma_1 x) + (1-\lambda)\sigma_2^2 \exp(-2\sigma_2 x)]^{1/2},$$

where λ is determined from the condition

$$D\zeta(p_{12}^*, \sigma_1) = D\zeta(p_{12}^*, \sigma_2) = D_{12}^*(\sigma_1, \sigma_2), \quad 0 \le \lambda \le 1.$$

Calculations were used to verify the inequality

$$D\zeta(p_{12}^*, \sigma) \le D_{12}^*(\sigma_1, \sigma_2), \quad \sigma_1 \le \sigma \le \sigma_2.$$

According to the remark to Lemma C.1 this Lemma makes it possible for p_{12}^* to solve the minimax problem on the segment $[\sigma_1, \sigma_2]$. However, *we have no strict proof of the sought inequality.*

In addition, let us note that here practically the density

$$p^{(0)}(t) = [\sigma_1 \exp(-\sigma_1 t) + \sigma_2 \exp(-\sigma_2 t)]/2$$

is almost a minimax one, i.e. $F_\zeta(p^{(0)}) \approx F_\zeta(p^*)$ (see [49], Sect. 4.3). This density is convenient for the 'method of similar trajectories', in which $\|K_p\| = C\, F_\zeta(p)$ ([49], Sect. 4.3).

References

[1] G.E. Albert, A general theory of stochastic estimates of the Neumann series for the solution of certain Fredholm integral equations and related series. In: *Symposium on Monte Carlo methods* (ed. H.A. Meyer) John Wiley, New York, 1956, pp. 37-46.

[2] H. Amann, Optimale Anfangsverteilungen bei der Monte-Carlo-Methode mit Informationsspeicherung, *ZAMM* (1967) **47**, No. 5, 285–299 (in German).

[3] T.W. Andersson, *Introduction to Multivariate Statistical Analysis.* John Wiley, New York, 1958.

[4] V.Ya. Arsenin, *Methods of Mathematical Physics and Special Functions.* Nauka, Moscow, 1974 (in Russian).

[5] N.S. Bakhvalov, *Numerical Methods.* Nauka, Moscow, 1973 (in Russian).

[6] A.A. Borovkov, *Probability Theory.* Nauka, Moscow, 1986 (in Russian).

[7] Yu.V. Bulavsky, On choosing the survival factor in the Neumann-Ulam scheme. In: *Methods and Algorithms of Statistical Modelling*, Comp. Cent. Sib. Branch, USSR Acad. Sci., Novosibirsk, 1983, pp. 13–20 (in Russian).

[8] Yu.V. Bulavsky. Minimization of collision estimate variance in problems with alternating sign terms. *Zh. Vychisl. Mat. Mat. Fiz.* (1984) **24**, No. 1, 148–152 (in Russian).

[9] Yu.V. Bulavsky and S.A. Temnikov, Randomized method of successive approximations. In: *Mathematical methods in Stochastic Simulation and Experimental Design*, St. Petersburg University Publishing House, 1996, pp. 64–68.

[10] V.I. Bykov and G.S. Yablonskii, One simplest catalytic reaction. *React. Kinet. Catal. Lett.* (1981) **16**, No. 7, 377-381.

[11] R. Courant, *Partial Differential Equations.* Mir, Moscow, 1964 (Russian translation).

[12] J.H. Curtiss. Monte Carlo methods for the iteration of linear operators. *J. Math. Phys.* (1954) **32**, 209–232.

[13] N. Danford and J.T. Schwartz, *Linear Operators. Part I. General Theory.* Interscience Publishers, New York, 1958.

[14] B. Davison, *Neutron Transport Theory.* Oxford University Press, 1957.

[15] B.S. Elepov, A.A. Kronberg, G.A. Mikhailov, and K.K. Sabelfeld, *Solution of Boundary-Value Problems by the Monte Carlo Method.* Nauka, Novosibirsk, 1980 (in Russian).

[16] B.S. Elepov and G.A. Mikhailov, On the solution of the Dirichlet problem for equation $\Delta u + cu = -g$ by the 'walk on spheres' method. *Zh. Vychisl. Mat. Mat. Fiz.* (1969) **9**, No. 3, 647–654 (in Russian).

[17] B.S. Elepov and G.A. Mikhailov, Using fundamental solutions to elliptic equations for constructing Monte Carlo algorithms. *Zh. Vychisl. Mat. Mat. Fiz.* (1974) **14**, No. 3, 728–736 (in Russian).

[18] S.M. Ermakov. The Monte Carlo method for iterating nonlinear operators. *Dok. Akad. Nauk SSSR* (1972) **204**, No. 2, 271–274 (in Russian).

[19] S.M. Ermakov and G.A. Mikhailov, *Statistical Modelling*. Nauka, Moscow, 1982 (in Russian).

[20] S.M. Ermakov, V.V. Nekrutkin and A.S. Sipin. *On the Use of Stochastic Processes for Solving Classical Equations of Mathematical Physics*. Nauka, Moscow, 1984 (in Russian).

[21] S.M. Ermakov, V.G. Zolotukhin, Monte Carlo applications to the nuclear shielding problem. In: *Voprosy Fiz. Zazc. React.* (ed. D.L. Broder) Atomizdat, Moscow, 1963, pp. 171-182 (in Russian)

[22] W. Feller, *An Introduction to Probability Theory and its Application, Vol.2*. John Wiley, New York, 1971.

[23] J.N. Franclin, Deterministic simulation of random processes. *Math. Comp.* (1963) **17**, 28-59.

[24] A.D. Frank-Kamenetskii and M.S. Judkevich, Monte Carlo calculation of the life time of momentary neutrons in a nuclear reactor. *Preprint No. 2155*, INE, Moscow, 1971 (in Russian).

[25] M. Freidlin, *Functional Integration and Partial Differential Equations*. Princeton University Press, Princeton, 1985 (*Ann. of Math. Stud.*, **109**).

[26] A.S. Frolov and N.N. Chentsov, On Monte Carlo calculation of definite integrals dependent on a parameter. *Zh. Vychisl. Mat. Mat. Fiz.* (1962) **2**, No. 4, 714–717 (in Russian).

[27] A.V. Grechannikov, Solution of Problems of Fluctuating Chemistry by Monte Carlo method. *Thes. Doct. Diss.*, Novosibirsk, 1983 (in Russian).

[28] J.M. Hammersley and K.W. Morton, A new Monte Carlo technique–antithetic variates, *Proc. Cambr. Phil. Soc.* (1956) **52**, 449-474.

[29] P. Hanusse, Monte Carlo simulations of finite fluctuations. *J. Chem. Phys.* (1977) **67**, 1282-1284.

[30] J.H. Halton, A retrospective and prospective survey of the Monte Carlo methods, *SIAM Rev.*, (1970) **12**, 1–63.

[31] M. Kac, *Probability and Related Topics in Physical Sciences*. Mir, Moscow, 1965 (Russian translation).

[32] Kai Lai Chzhun, *Homogeneous Markov Chains*. Mir, Moscow, 1964 (in Russian).

[33] M.H. Kalos and P.A. Whitlock, *Monte Carlo Methods. Vol. 1: Basics*. John Wiley, New York, 1986.

[34] O.B. Kamaeva Trace length estimate in the weight Monte Carlo methods. *Zh. Vychisl. Mat. Mat. Fiz.*, (1984) **24**, No. 4, 514-519 (in Russian).

[35] A.A. Kronberg, On algorithms for statistical modelling of the solution of boundary-value problems of elliptic type. *Zh. Vychisl. Mat. Mat. Fiz.* (1984) **24**, No. 10, 1531-1537 (in Russian).

[36] T.A. Kurtz, The relationship between stochastic and deterministic models for chemical reactions. *J. Chem. Phys.* (1972) **57**, 2976-2978.

[37] O.A. Ladyzhenskaya, V.A. Solonnikov, and N.N. Ural'tseva, *Linear and Quasilinear Equations of Parabolic Type*, Nauka, Moscow, 1967, (in Russian).

[38] G.Z. Lotova, Investigation of the transfer process in a stochastic medium by calculating the derivatives with respect to a random parameter. In: *Proceedings of the Third St.-Petersburg Workshop on Simulation*, St. Petersburg University Publishing House, 1998, pp. 89-94.

[39] R.N. Makarov, Investigation of Monte Carlo estimates for solving the nonlinear Helmholtz equation. In: *Vichisl. matem.*, No. 4, Computing center, Novosibirsk, 1996 (in Russian).

[40] G.I. Marchuk, *Methods of Nuclear Reactor Calculation*. Gosatomizdat, Moscow, 1961 (in Russian).

[41] G.I. Marchuk, G.A. Mikhailov, M.A. Nazaraliev, *et al. The Monte Carlo Method in Atmospheric Optics*. Springer, Berlin–Heidelberg, 1980.

[42] Matsushima, Model chemical reactions with several steady states, *Bull. Chem. Soc. Japan* (1978) **51**, No. 7, 1956-1960.

[43] D.McQuarrie, Stochastic approach to chemical kinetics. *Suppl. Rev. Ser. Appl. Probl.*, Methuen, London, 1967.

[44] G.A. Mikhailov, A method of simulating the free path of a particle. *Atomn. Energiya* (1970) **28**, No. 2, 175 (in Russian).

[45] G.A. Mikhailov, A new Monte Carlo algorithm for estimating the maximum eigenvalue of an integral operator. *Docl. Akad. Nauk SSSR* (1970) **191**, No. 5, 993–996 (in Russian).

[46] G.A. Mikhailov, Estimation of the cost of simulating the process of random walk by spheres for some types of regions. *Zh. Vychisl. Mat. Mat. Fiz.* (1979) **19**, No. 2, 510–515.

[47] G.A. Mikhailov, Asymptotics of the average radiation intensity for some models of stochastic media. *Izv. Akad. Nauk SSSR, Fiz. Atmosf. Okeana* (1982) **18**, No. 12, 1289–1295 (in Russian).

[48] G.A. Mikhailov, *Minimization of Computational Costs of Nonanalogue Monte Carlo Methods*. World Scientific, Singapore, 1991.

[49] G.A. Mikhailov, *Optimization of Weighted Monte Carlo Methods*. Springer, Berlin–Heidelberg, 1992.

[50] G.A. Mikhailov,*New Monte Carlo Methods with Estimating Derivatives*. VSP, Utrecht, 1995.

[51] G.A. Mikhailov and M.V. Antipov, Estimates of distribution nonuniformity of congruence sums of random variables. *Dokl. Rus. Akad. Nauk* (1996) **347**, No. 1, 23–26 (in Russian).

[52] G.A. Mikhailov and G.Z. Lotova, Estimating probabilistic moments of critical parameters of the equation of particle transfer in stochastic medium. *Dokl. Rus. Akad. Nauk* (1997) **356**, No. 2, 166–169 (in Russian).

[53] G.A. Mikhailov and R.N. Makarov, Solution of boundary-value problems by 'walk on spheres' method with reflection from the boundary. *Dokl. Rus. Akad. Nauk* (1997) **353**, No. 6, 720–722 (in Russian).

[54] G.A. Mikhailov and B.V. Men'shchikov, Solution of the Helmholtz equation with a complex parameter and realization of the Fourier transformation by the Monte Carlo method. *Dokl. Rus. Akad. Nauk* (1996) **349**, No. 1, 17–19 (in Russian).

[55] G.A. Mikhailov and M.Ju. Plotnikov, An estimator 'over a run' for a solution of linear and nonlinear radiative transfer equations in the large. *Dokl. Rus. Akad. Nauk* (1994) **337**, No. 2, 162–164.

[56] G.A. Mikhailov and A.S. Serednyakov, Numerical and asymptotic estimates of a model dimension influence on radiation transfer. *Optika atmosf. okeana* (1997) **10**, No. 3, 201–210 (in Russian).

[57] *Monte Carlo Methods in Statistical Physics* (ed. K. Binder) Springer-Verlag, Berlin, 1979.

[58] M. Motoo, Some evaluations for the continuous Monte Carlo method by using the Brownian hitting process. *Ann. Stat. Math.* (1959) **11**, 49–54.

[59] M.E. Muller, Some continuous Monte Carlo methods for the Dirichlet problem. *Ann. Math. Stat.* (1956) **27**, No. 3, 569–589.

[60] G.Nicolis and I. Prigogine, *Self-Organization in Nonequilibrium Systems.* Wiley, New York, 1979.

[61] M.Yu. Plotnikov. Using the weighted Monte Carlo method for solving nonlinear integral equations. *Russ. J. Numer. Anal. Math. Modelling* (1994) **9**, No. 2, 1–24.

[62] S.I. Pokhozhayev. On the Dirichlet problem for the equation $\Delta u = u^2$. *Dokl. Akad. Nauk SSSR* (1960) **134**, No. 4, 769–773 (in Russian).

[63] Yu.V. Prokhorov and Yu.A. Rosanov, *Theory of Chances.* Nauka, Moscow, 1973 (in Russian).

[64] A.P. Prudnikov, Yu.A. Brychkov, and O.I. Marichev, *Integrals and Series.* Nauka, Moscow, 1981 (in Russian).

[65] R.T. Rockafellar. *Convex Analysis.* Princeton University Press, Princeton, NJ, 1970.

[66] Yu.A. Romanov, Exact solutions of a one-speed kinetic equation and their use in the calculation of diffusion problems (improved diffusion method). In: *Study of the Critical Parameters of Reactor Systems.* Gosatomizdat, Moscow, 1960, pp. 3–26 (in Russian).

[67] I.A. Shalimova, Spherical mean-value theorem at a noncentral point for the Lamé system and its use when constructing Monte Carlo algorithms. *Rus. J. Numer. Anal. Math. Modelling* (1993) **8**, No. 4, 327–337.

[68] D.I. Sharikalov and S.A. Uhinov, Monte Carlo estimates for derivatives of polarized radiation. In: *Proceedings of the Third St.-Petersburg Workshop on Simulation,* St. Petersburg University Publishing House, 1998, pp. 126-131.

[69] A.F. Sizova. On the variance of collision estimate for solving nonlinear problems by Monte Carlo method. *Vestnik Leningradskogo Univ.* (1976), No.1, 152–155 (in Russian).

[70] I.M. Sobol, On one approach to the calculation of multidimensional integrals. *Problems of Computational and Applied Mathematics* (1970) **38**, 100–111, Tashkent (in Russian).

[71] J. Spanier and E.M. Gelbard, *Monte Carlo Principles and Neutron Transport Problems.* Addison-Wesley, Reading, PA, 1969.

[72] K.D. Tocher, The application of automatic computers to sampling experiments. *J. Royal Statist. Soc. Ser.* B (1954) **16**, No. 1, 39–61.

[73] J.S. Turner, Finite fluctuations, nonlinear thermodynamics and far-from equilibrium transitions between multiple steady states. *Adv. in Chem. Phys.* (1975) **29**, 63–78.

[74] A.D. Ventzel and M.I. Freydlin, *Fluctuations in Dynamic Systems under Action of Small Random Perturbations.* Nauka, Moscow, 1979 (in Russian).

[75] A. Wald, Statistical decision functions. *Ann. Math. Stat.* (1949) No. 20.

[76] G.S. Yablonskii and S.I. Spivak, *Mathematical models of chemical kinetics.* Moscow, 1977 (in Russian).

[77] V.G. Zolotukhin and L.V. Maiorov, *Monte Carlo Assessment of Criticality Parameters of Reactors.* Energoatomizdat, Moscow, 1984 (in Russian).

Bei Fragen zur Produktsicherheit wenden Sie sich bitte an:
If you have any questions regarding product safety,
please contact:

Walter de Gruyter GmbH
Genthiner Straße 13
10785 Berlin
productsafety@degruyterbrill.com